GCSE Success

REVISION GUIDE

Additional Science

Carol Tear
Emma Poole
Ian Honeysett

Contents

Biology

Revised

4	Cells and Organisation	
6	DNA and Protein Synthesis	
8	Proteins and Enzymes	
10	Cell Division	
12	Growth and Development	
14	Transport in Cells	
16	Respiration	
18	Sampling Organisms	
20	Photosynthesis	
22	Food Production	
24	Transport in Animals	
26	Transport in Plants	
28	Digestion and Absorption	
30	Practice Questions	

Physics

Revised

32	Distance, Speed and Velocity	
34	Speed, Velocity and Acceleration	
36	Forces	
38	Acceleration and Momentum	
40	Pairs of Forces: Action and Reaction	
42	Work and Energy	
44	Energy and Power	
46	Electrostatic Effects	
48	Uses of Electrostatics	
50	Electric Circuits	
52	Voltage or Potential Difference	
54	Resistance and Resistors	
56	Special Resistors	
58	The Mains Supply	
60	Atomic Structure	
62	Radioactive Decay	
64	Living with Radioactivity	
66	Uses of Radioactive Material	
68	Nuclear Fission and Fusion	
70	Stars	
72	Practice Questions	

Chemistry

Revised

- 74 Atomic Structure
- 76 Atoms and the Periodic Table
- 78 The Periodic Table
- 80 Chemical Reactions and Atoms
- 82 Balancing Equations
- 84 Ionic and Covalent Bonding
- 86 Ionic and Covalent Structures
- 88 Group 7
- 90 New Chemicals and Materials
- 92 Plastics and Perfumes
- 94 Analysis
- 96 Metals
- 98 Group 1
- 100 Aluminium and Transition Metals
- 102 Chemical Tests
- 104 Acids and Bases
- 106 Making Salts
- 108 Metal Carbonate Reactions
- 110 The Electrolysis of Sodium Chloride Solution
- 112 Relative Formula Mass and Percentage Composition
- 114 Calculating Masses
- 116 Rates of Reaction
- 118 Reversible Reactions
- 120 Energy Changes
- 122 Practice Questions

124 Answers

128 Index

A copy of the periodic table can be found on the inside back cover.

Cells and Organisation

Plant and Animal Cells

Plants and animal cells have a number of structures in common.

They all have:
- A **nucleus** that carries genetic information and controls the cell.
- A **cell membrane** which controls the movement of substances in and out of the cell.
- **Cytoplasm** where most of the chemical reactions happen.

There are three main differences between plant and animal cells:
- Plant cells have a strong **cell wall** made of cellulose, whereas animal cells do not. The cell wall supports the cell and stops it bursting.
- Plant cells have a large permanent **vacuole** containing cell sap, but vacuoles in animal cells are small and temporary. The cell sap is under pressure and this supports the plant.
- Plant cells may contain **chloroplasts** containing chlorophyll for photosynthesis. Animal cells never contain chloroplasts.

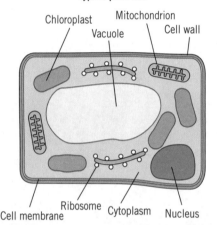

Typical plant cell

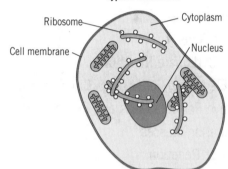

Typical animal cell

Build Your Understanding

The naked eye can see detail down to about 0.1 mm. Cells are smaller than this so a microscope is needed to see individual cells.

The best light microscopes can magnify cells so that objects as small as 0.002 mm can be seen clearly. At this magnification other structures in the cell become visible but cannot be seen clearly. Using an electron microscope allows objects as small as 0.000002 mm to be seen:
- Mitochondria are the site of respiration in the cell.
- Ribosomes are small structures in the cytoplasm where proteins are made.

Becoming larger and multicellular does have some advantages but also produces difficulties:
- It allows cells to specialise. This makes them more efficient at their job.

The difficulties that need to be solved are:
- It requires a communication system between cells to be developed.
- It is harder to supply all the cells with nutrients.
- The surface area to volume ratio is smaller so it is harder to exchange substances with the environment.

✓ Maximise Your Marks

To get an A* you must be able to measure cells and structures from a drawing and then use the magnification of the drawing to work out their size in real life. An example of this is question 1 in the Stretch Yourself section. Remember that magnification = image size ÷ size in real life.

Levels of Organisation

Some organisms are made of one cell. They are **unicellular**.

There seems to be a limit to the size of a single cell so larger organisms are made up of a number of cells. They are **multicellular**.

The cells are not all alike, but are specialised for particular jobs, for example guard cells in leaves and neurones in animals.

Similar cells, doing similar jobs, are gathered together into **tissues**, such as xylem and nerves. Different tissues are gathered together into **organs** to do a particular job, for example leaves and the brain.

Groups of organs often work together in **systems** to carry out related functions.

Organs work together in systems

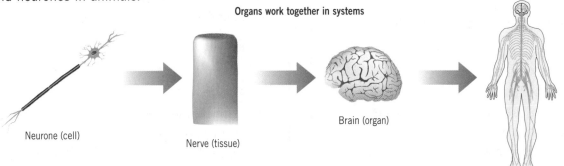

Neurone (cell) → Nerve (tissue) → Brain (organ) → Nervous system (organ system)

💡 Boost Your Memory

Make sure that you can remember the order of terms to describe the levels of organisation: **c**ells, **t**issues, **o**rgans, **s**ystems. Think of a way of remembering this.

✓ Maximise Your Marks

Be careful with 'bone' and 'muscle'. Bone and muscle are both tissues. However a bone and a muscle are both organs as they contain a number of different tissues.

Structure of Bacteria

Bacterial cells are very different from plant and animal cells. That is why they are classified in a different kingdom (Prokaryotes).

They vary in shape, but all bacterial cells have a similar structure.

Bacterial cells are smaller than animal and plant cells.

They lack a true nucleus, mitochondria, chloroplasts and vacuoles.

❓ Test Yourself

1. Write down three structures that are found in plant cells but not in animal cells.
2. What is the main function of cell sap?
3. At what level of organisation are leaves and the brain?

⭐ Stretch Yourself

1. The nucleus in the animal cell shown on page 4 is 0.005 mm wide in real life. What is the magnification of this diagram?
2. Specialisation allows cells to become more efficient at their particular role. A disadvantage of specialisation is that an organism cannot live if it loses certain organs. Suggest why this is.

DNA and Protein Synthesis

The Structure of DNA

The nucleus controls the chemical reactions occurring in the cell. This is because it contains the genetic material.

This is contained in structures called **chromosomes** which are made of **DNA**.

DNA is a large molecule with a very important structure:
- It has two stands.
- The stands are twisted to make a shape called a **double helix**.
- Each strand is made of a long chain of molecules including sugars, phosphates and bases.
- There are only four bases, called A, C, G and T.
- Hydrogen bonds between the bases, hold the two chains together. C always links with G, and A with T.

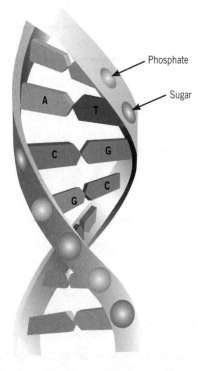

Boost Your Memory

You need to remember **A** with **T** and **C** with **G**. Find a way of remembering it that means something to you. It could be **A**untie **T**ina and **C**ousin **G**eorge, for example.

Discovering the Structure of DNA

The two scientists that are famous for discovering the structure of DNA are Francis Crick and James Watson. They worked together in Cambridge in the early 1950s.

A molecule of DNA is about 0.00000034 mm wide, so they could not use a microscope to see it.

This is where they needed information obtained by other scientists:
- Maurice Wilkins and Rosalind Franklin fired X-rays at DNA crystals, and the images they obtained told Watson and Crick that DNA was shaped like a helix, with two chains.
- Erwin Chargaff had worked out that there was always the same percentage of the base C as G, and the same percentage of A as T.

These two pieces of information allowed Watson and Crick to build their famous model of the structure.

✓ Maximise Your Marks

How Science Works questions may ask for examples of how advances in science are made by cooperation between scientists. The discovery of the structure of DNA is a good example to use.

Build Your Understanding

DNA codes for the proteins it makes are called bases. There are four bases:
- adrenine (abbreviated A)
- cytosine (C)
- guanine (G)
- thymine (T)

They are given the letters A, C, G and T.

Coding for Proteins

DNA controls the cell by carrying the code for proteins:
- Each different protein is made of a particular order of amino acids, so DNA must code for this order.
- A gene is a length of DNA that codes for the order of amino acids in one protein.

Scientists now know that each amino acid in a protein is coded for by each set of three bases along the DNA molecule.

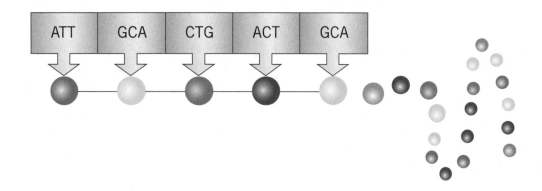

Build Your Understanding

Proteins are made on ribosomes in the cytoplasm and DNA is kept in the nucleus and cannot leave.

The cell has to use a messenger molecule to copy the message from DNA and carry the code to the ribosomes.

This molecule is called messenger RNA (mRNA).

When a protein is to be made, these steps occur:
- The DNA containing that gene unwinds and 'unzips'.
- Complementary mRNA molecules pair up next to the DNA bases on one strand.
- The mRNA units join together and make a molecule with a complementary copy of the gene. This is called transcription.
- The mRNA molecule then leaves the nucleus and attaches to a ribosome.
- The base code on the mRNA is then used to link amino acids together in the correct order to produce the protein. Each three bases code for one amino acid. This is called translation.

✓ Maximise Your Marks

The word 'complementary' is a useful word to use when describing mRNA. It does not copy the code exactly but makes a version using the matching bases.

❓ Test Yourself

1. Why is the shape of DNA described as a double helix?
2. How does the structure of one protein differ from the structure of another protein?
3. What holds the two strands together in a DNA molecule?
4. Why is the genetic code described as a triplet code?

⭐ Stretch Yourself

1. In a length of DNA, 34% of the bases are the base G. What percentage are base T?
2. When a protein is to be made, the length of DNA containing that gene 'unzips'. What does this mean and why is it necessary?

Proteins and Enzymes

The Functions of Proteins

The only way that the genetic material can control the cell is by coding for which proteins are made.

The proteins that are produced have a wide range of different functions:
- Structural proteins used to build cells, e.g. collagen.
- Hormones to carry messages, e.g. insulin.
- Carrier molecules, e.g. haemoglobin.
- Enzymes to speed up reactions, e.g. amylase.

Enzymes

Enzymes are **biological catalysts**.

Enzymes are produced in all living organisms and control all the chemical reactions that occur.

Most of the chemical reactions that occur in living organisms would occur too slowly without enzymes. Increased temperatures would speed up the reactions, but using enzymes means that the reactions are fast enough at 37°C.

These reactions include DNA replication, digestion, photosynthesis, respiration and protein synthesis.

✓ Maximise Your Marks

Many people think that all enzymes are released into the gut to digest food. Remember that most enzymes are found inside cells and are not released.

How do Enzymes Work?

As enzymes are protein molecules, they are made of a long chain of amino acids that is folded up to make a particular shape.

They have a slot or a groove, called the **active site**, into which the **substrate** fits. The substrate is the substance that is going to react.

The reaction then takes place and the **products** leave the enzyme. This explanation for how enzymes work is called the **Lock and Key theory**.

The substrate fits into the active site like a key fitting into a lock.

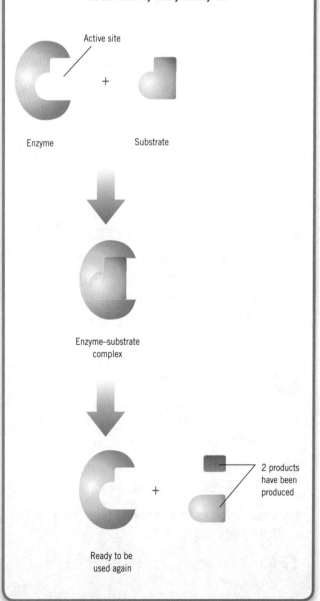

The lock and key theory of enzymes

What Factors Affect Enzymes?

Enzymes work best at a particular temperature and pH. This is called the **optimum** temperature or pH.

Enzymes have different optimum values that depend on where they usually work.

If the concentration of the substrate is increased, then the reaction will be faster up to a certain concentration and then it will level off until all the enzymes are working at their maximum rate. At this point, increasing the substrate concentration does not increase the rate of reaction.

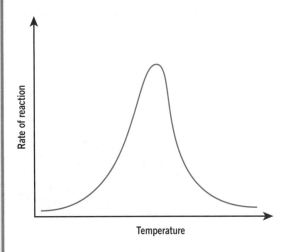

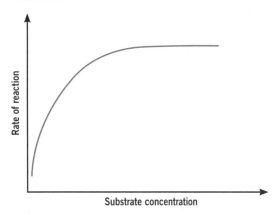

Build Your Understanding

The Lock and Key theory can be used to explain many of the properties of enzymes:
- It explains why an enzyme will only work on one type of substrate. They are described as specific. The substrate has to be the right shape to fit into the active site.
- If the temperature is too low, then the substrate and the enzyme molecules will not collide so often and the reaction will slow down. If the shape of the enzyme molecule changes, then the substrate will not easily fit into the active site. This means that the reaction will slow down.

High temperatures and extremes of pH may cause this to happen.

If the shape of the enzyme molecule is irreversibly changed, then it is described as being denatured.

✓ Maximise Your Marks

Many candidates lose marks by saying that heat kills enzymes. Remember that enzymes are protein molecules and not living organisms. Say that they are denatured or destroyed, but not killed.

❓ Test Yourself

1. Why does a lack of protein stunt growth?
2. Why are enzymes necessary in living organisms?
3. What is the lock and what is the key in the Lock and Key theory?
4. What does the phrase 'optimum temperature' mean?

⭐ Stretch Yourself

1. Lipase digests fats, but it will not digest proteins. Explain why this is.
2. Adding vinegar to food can stop the food being digested and spoilt by bacteria and fungi. Explain why this is.

Cell Division

Copying the DNA

Before a cell divides, two things must happen. Firstly, new cell organelles such as mitochondria must be made. Secondly, the DNA must copy itself. Watson and Crick realised that the structure of DNA allows this to happen in a rather neat way:

- The double helix of DNA unwinds and the two strands come apart or 'unzip'.
- The bases on each strand attract their complementary bases and so two new molecules are made.

DNA replication

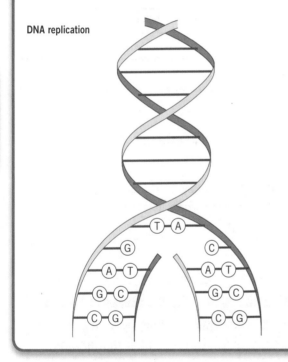

Types of Cell Division

Cells divide for a number of reasons. There are two types of cell division – **meiosis** and **mitosis** – and they are used for different reasons.

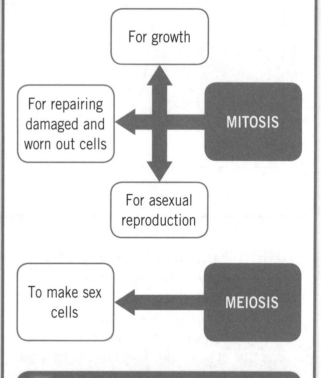

✓ Maximise Your Marks

In most questions you will not lose marks if your spelling is a little inaccurate. However, make sure that you spell 'mitosis' and 'meiosis' correctly or the examiner may not know which one you mean.

Mitosis

In **mitosis**, two cells are produced from one. As long as the chromosomes have been copied exactly, then both new cells will have the same number of chromosomes and therefore the same genetic information as each other and the parent cell.

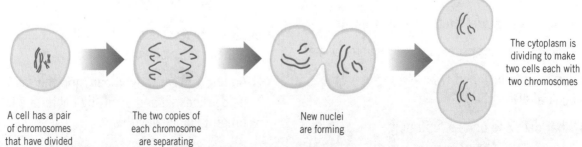

A cell has a pair of chromosomes that have divided → The two copies of each chromosome are separating → New nuclei are forming → The cytoplasm is dividing to make two cells each with two chromosomes

Meiosis

In **meiosis**, the chromosomes are also copied once, but the cell divides twice. This makes four cells each with half the number of chromosomes, one from each pair.

Cells with one chromosome from each pair are called **haploid** and can be used as **gametes**. When two gametes join at fertilisation, the **diploid** or full number of chromosomes is produced.

A cell has a pair of chromosomes each of which has divided

The two chromosomes are separating

Each double stranded chromosome is now split up

Four new cells are formed each with one chromosome

Build Your Understanding

When DNA is copied, before mitosis and meiosis occur, mistakes are sometimes made.

A **gene mutation** occurs when one of the chemical bases in DNA is changed. This may mean that a different amino acid is coded for and this can change the protein that is made.

When this happens, it is most unlikely to benefit the organism. Either the protein will not be made at all or most likely it will not work properly.

Very occasionally, a mutation may be useful, and without mutations we would not be here.

Mutations occur randomly at a very low rate, but some factors can make them happen more often. These include:
- Ultra-violet light in sunlight.
- X-rays.
- Chemical mutagens as found in cigarettes.

Only mutations can produce new genes, but meiosis can recombine them in different orders.

Also, as a baby can receive any one of the chromosomes in each pair from the mother and any one from the father, the number of possible gene combinations is enormous.

This new mixture of genetic information produces a great deal of variation in the offspring.

This is why meiosis and sexual reproduction produces so much more variation than asexual reproduction.

✓ Maximise Your Marks

Remember that mitosis can produce cells that are genetically different, but this only happens if there is a mutation. Otherwise, they are genetically identical. Meiosis always produces genetic variation.

❓ Test Yourself

1. Does a new molecule of DNA have none, one or two original strands?
2. Where in the human body does meiosis occur?
3. The haploid number of chromosomes in humans is 23. What is the diploid number?
4. Write down two differences between mitosis and meiosis.

⭐ Stretch Yourself

1. Why is a gene mutation often harmful?
2. Why is it important to make sure that your sunglasses filter out UV light?

Growth and Development

Division and Differentiation

When gametes join at fertilisation, this produces a single cell called a **zygote**. The zygote soon starts to divide many times by mitosis to produce many identical cells.

These cells then start to become specialised for different jobs. The production of different types of cells for different jobs is called **differentiation**. These differentiated cells can then form tissues and organs.

Stem Cells

Some cells in the embryo and in the adult keep the ability to form other types of cells. They are called **stem cells**.

Scientists are now trying to use stem cells to replace cells that have stopped working or been damaged. This has the potential to cure a number of diseases.

Build Your Understanding

Once a cell has differentiated, it does not form other types of cell.

Although it has the same genes as all the other cells, many are turned off so it only makes the proteins it needs.

Scientists have found a way to switch genes back on and so have been able to clone animals from body cells.

This means that it is now possible to produce embryos that are clones of an animal and to use them to supply embryonic stem cells.

There are many different views about the possibility of cloning humans to obtain stem cells.

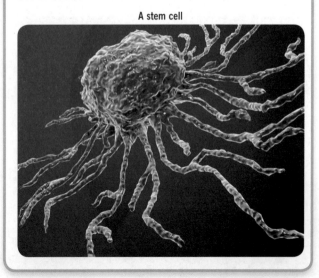

A stem cell

Human Growth Curves

Humans grow at different rates at different stages of their lives. This is shown in the graph.

The graph shows that there are two phases of rapid growth, one just after birth and the other in puberty.

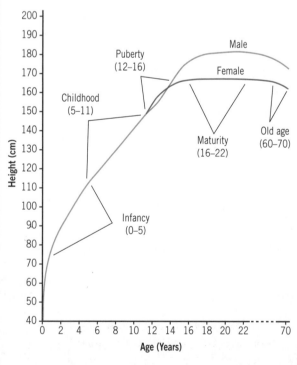

Human growth at different ages

Build Your Understanding

The various parts of the body also grow at different rates at different times.

The diagram shows that the head and brain of an early foetus grow very quickly compared with the rest of the body.

Later, the body and legs start to grow faster while the brain and head grow more slowly into puberty and adulthood.

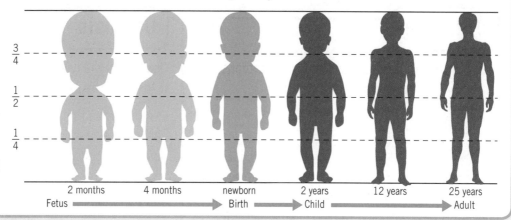

Plant Growth

Like animals, plants grow by making new cells by mitosis. The cells then differentiate into tissues like xylem and phloem. These tissues then form organs such as roots, leaves and flowers.

Growth in plants is different from animal growth in a number of ways:
- Plant cells enlarge much more than animal cells after they are produced. This increases the size of the plant.
- Cells tend to divide at the ends of roots and shoots. This means that plants grow from their tips.
- Animals usually stop growing when they reach a certain size, but plants carry on growing.
- Many plant cells keep the ability to produce new types of cells, but in animals only stem cells can do this. Plant cells that can produce new types of cells are called **meristematic**.

Measuring Growth

Growth can be measured as an increase in **height**, **wet mass** or **dry mass**. Dry mass is the best measure of growth. There are advantages and disadvantages of measuring growth by each method.

Measurement	Advantage	Disadvantage
Length or height	Easy to measure	Only measures growth in one direction
Wet mass	Fairly easy to measure	Water content can vary
Dry mass	Measures permanent growth over the whole body	Involves removing all the water from an organism

✓ Maximise Your Marks

It is easy to use the word 'weight' when talking about measuring growth, but you should really say 'wet mass' or 'dry mass'.

❓ Test Yourself

1. Write down one type of specialised cell.
2. What is a stem cell?
3. Look at the human growth curve. What does it show about growth in old age?
4. Which parts of a plant contain the main growth areas?

⭐ Stretch Yourself

1. Suggest why the head is much larger than the rest of the body when the baby is young?
2. Using dry mass to measure the growth of an organism presents a number of difficulties. Suggest what these difficulties are.

Transport in Cells

Diffusion

Substances can pass across the cell membrane by three different processes:
- diffusion
- osmosis
- active transport.

Diffusion is the movement of a substance from an area of high concentration to an area of low concentration. Diffusion works because particles are always moving about in a random way. The rate of diffusion can be increased in a number of ways:

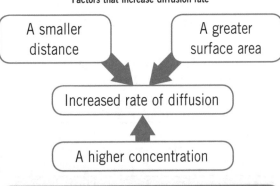

Factors that increase diffusion rate

💡 Boost Your Memory

You must remember than diffusion is high to low concentration. Perhaps the letters DHL might help you remember this?

Osmosis

Osmosis is really a special type of diffusion. It involves the diffusion of water.

Osmosis is the movement of water across a partially permeable membrane from an area of high water concentration to an area of low water concentration.

The cell membrane is an example of a partially permeable membrane. It lets small molecules through, such as water, but stops larger molecules, such as glucose.

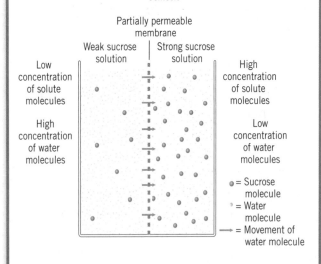

Active Transport

Sometimes substances have to be moved from a place where they are in low concentration to one where they are in high concentration. This is in the opposite direction to diffusion and is called **active transport**.

Active transport is therefore the movement of a substance against the diffusion gradient with the use of energy from respiration.

Anything that stops respiration will therefore stop active transport. For example, plants take up minerals by active transport. Farmers try and make sure that their soil is not waterlogged because this reduces the oxygen content of the soil, so less oxygen is available to the root cells for respiration. This would therefore reduce the uptake of minerals.

✓ Maximise Your Marks

You can use the phrase 'up' or 'against the diffusion gradient' because this means in the opposite direction to diffusion. Don't say that active transport is 'along' or 'down the diffusion gradient' because this is the wrong way.

Build Your Understanding

When plant cells gain water by osmosis, they swell. The cell wall stops them from bursting. This makes the cell stiff or turgid.

If a plant cell loses water it goes limp or flaccid.

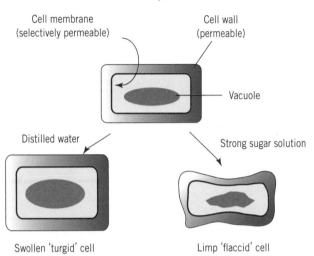

Osmosis in plant cells

Turgid cells are very important in helping to support plants. Plants with flaccid cells often wilt.

Sometimes the cells lose so much water that the cell membrane may come away from the cell wall. This is called plasmolysis.

Animal cells do not behave in the same way because they do not have a cell wall.

They will either swell up and burst if they gain water, or shrink if they lose water.

It is possible to show how osmosis has occurred by cutting cylinders out of a potato and putting them into sugar solutions of different concentrations.

If the mass of the chips is measured before and after they are put in the solutions, a graph like this can be plotted:

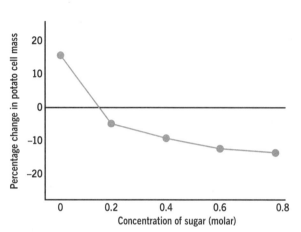

In less concentrated solutions, the potato gains water and increases in mass.

At high concentrations of sugar, the potato loses water and decreases in mass.

✓ Maximise Your Marks

Make sure you can explain the shape of this graph. It might be drawn as the length of the potato cylinder rather than the mass. It's just the same.

❓ Test Yourself

1. Why does the smell from a 'stink bomb' spread across a room?
2. Why do bad smells travel faster in warmer weather?
3. What is a partially permeable membrane?
4. What substance moves in osmosis?

⭐ Stretch Yourself

1. Look at the graph showing the percentage change in mass of the potato chips. What concentration of sugar solution is equal to the concentration inside a normal potato?
2. Why does a waterlogged soil have less oxygen in it?

Respiration

What is energy needed for?

The energy that is released by respiration can be used for many processes:
- To make large molecules from smaller ones, for example proteins from amino acids.
- To contract the muscles.
- For mammals and birds to keep a constant temperature.
- For active transport.

Build Your Understanding

The energy released by respiration is needed for different processes in different parts of the cell. To make sure that the energy is not lost as heat, it is trapped in the bonds of a molecule called ATP.

ATP can then pass the energy on to wherever it is needed.

Aerobic Respiration

Aerobic respiration is when glucose reacts with oxygen to release energy. Carbon dioxide (CO_2) and water (H_2O) are released as waste products.

Glucose + Oxygen → Carbon dioxide + Water + Energy

$$C_6H_{12}O_6 + 6O_2 \rightarrow 6CO_2 + 6H_2O + \text{energy}$$

The reactions of aerobic respiration take place in mitochondria.

All the reactions that occur in our body are called our **metabolism** and so anything that increases our **metabolic rate** will increase our need for respiration.

During exercise the body needs more energy and so the rate of respiration increases.

The breathing rate increases to obtain extra oxygen and remove carbon dioxide from the body. The heart beats faster so that the blood can transport the oxygen, glucose and carbon dioxide faster.

This is why our pulse rate increases.

Boost Your Memory

When you are learning the equation for respiration, look at the equation for photosynthesis in the next topic (page 20). Remember that one is just the reverse of the other. Don't try to learn them separately.

Anaerobic Respiration

When not enough oxygen is available, glucose can be broken down by **anaerobic respiration**.

This may happen in muscle cells during hard exercise.

In humans:
Glucose → Lactic acid + Energy

Being able to respire without oxygen sounds a great idea. However, there are two problems:

- Anaerobic respiration releases much less energy than is released by aerobic respiration.
- Anaerobic respiration produces lactic acid which causes muscle fatigue and pain.

In plants and fungi, such as yeast, anaerobic respiration is often called **fermentation**. It produces different products.

In plants and fungi:
Glucose → Ethanol + Carbon dioxide + Energy

Build Your Understanding

The extra oxygen required after exercise to deal with the build up lactic acid is called oxygen debt. Another name for this is excess post-exercise oxygen consumption (EPOC).

The lactic acid is transported to the liver and the heart continues to beat faster to supply the liver with the oxygen needed to break down the lactic acid.

It is possible to measure the respiration rate of organisms by measuring:
- The oxygen consumption.
- The carbon dioxide production.

This apparatus opposite can be used to investigate the rate of oxygen consumption by the maggots. If a liquid that absorbs carbon dioxide is placed in the bottom of the test tube, then the coloured liquid will move to the left.

It is then possible to investigate the effect of factors such as temperature or pH on the rate of respiration, for example by carrying out the experiment at different temperatures.

It is also possible to calculate the respiratory quotient (RQ) using this formula:

$$RQ = \frac{\text{carbon dioxide produced}}{\text{oxygen used}}$$

The RQ provides useful information about what type of substance is being respired.

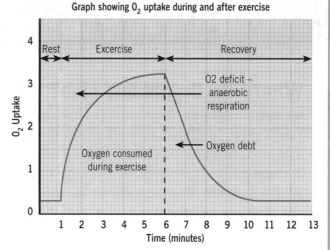

Graph showing O_2 uptake during and after exercise

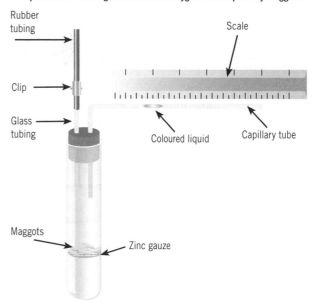

Experiment to investigation the rate of oxygen consumption by maggots

✓ Maximise Your Marks

Remember that respiration is controlled by enzymes. This means that any factors that change the rate of enzyme reaction will also change the rate of respiration.

❓ Test Yourself

1. Why do we need to eat more in cold weather?
2. Why do we breathe faster when we exercise?
3. What are the bubbles of gas given off when yeast is fermenting glucose?
4. Why do our muscles hurt when we run a long race?

⭐ Stretch Yourself

1. Look at the equation for aerobic respiration using glucose. What would be the RQ of an organism that is respiring only glucose?
2. Wine makers need to carefully control the temperature inside the fermentation tanks when they make wine using fermentation. Explain why.

Sampling Organisms

Where do Organisms Live?

Different organisms live in different environments.
- The place where an organism lives is called its **habitat**.
- All the organisms of one type living in a habitat are called a **population**.
- All the populations in a habitat are a **community**.
- An **ecosystem** is all the living and non-living things in a habitat.

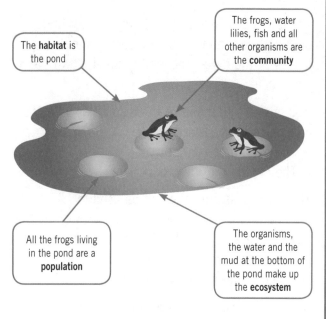

The **habitat** is the pond

The frogs, water lilies, fish and all other organisms are the **community**

All the frogs living in the pond are a **population**

The organisms, the water and the mud at the bottom of the pond make up the **ecosystem**

✓ Maximise Your Marks

Remember that for two organisms to be in the same population they must live in the same habitat and be in the same species. This means that they can successfully mate with each other.

Sampling an Area

It is possible to investigate where organisms live by using various devices.

A **quadrat** is usually a small square that is put on the ground within which all species of interest are noted or measurements taken. The number of organisms can be counted, or percentage cover estimated, in the quadrat and the size of the population in the whole area can then be estimated. Often several quadrats are required to determine the estimate.

It is easy to try to estimate how many of one type of plant live in a habitat:
- Work out the area of the whole habitat.
- Sample a small area using several quadrats and count out how many plants are present.
- Scale up this number to give an estimate of the whole habitat.

Quadrats are often used to study plants, but devices such as **pooters**, **nets** and **pit-fall traps** can be used to sample animal populations.

Working out the population of animals is harder because they do not keep still to be counted.

We can use a technique called **mark–recapture**:
- The organisms, such as snails, are captured, unharmed.
- They are counted and then marked in some way, for example the snail can be marked with a dot of non-toxic paint.
- They are released.
- Some time later the process of capturing is repeated and another count is made.
- This count includes the number of marked animals and the number of unmarked.

To work out the estimate of the population a formula is used. Population size is:

$$\frac{\text{number in 1st sample} \times \text{number in 2nd sample}}{\text{number in 2nd sample previously marked}}$$

✓ Maximise Your Marks

Remember in all sampling questions, the more samples that you take in an area, the more accurate the estimate of the whole area will be.

Mapping an Area

To estimate the size of a population in an area we can use quadrats put down at random.

To see where the organisms live in a habitat we can use a **transect line**:
- A tape measure (or a piece of string) is put down in a line across the habitat.
- Quadrats are put down at set intervals along the tape.
- The organisms in the quadrats are then counted.

Artificial Ecosystems

Our planet has a range of different ecosystems. Some of these are **natural**, such as woodland and lakes. Others are **artificial** and have been created by man, such as fish farms, greenhouses and fields of crops.

Artificial ecosystems usually have less variety of organisms living there (less biodiversity). This may be caused by the use of chemicals such as weedkillers, pesticides and fertilisers.

Build Your Understanding

In some habitats a transect line can produce interesting results.

Different organisms live at different points along the line. This is because there is a change in the environmental conditions along the line. This is called **zonation**.

An example of zonation is found in a pond. Different plants can grow at different distances into the pond. This is due to the amount of water in the soil.

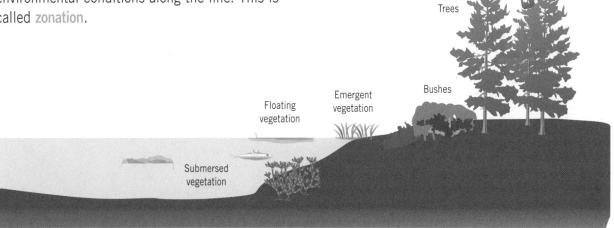

Test Yourself

1. What name is given to all the rabbits living in the same field?
2. What device would you use to sample:
   a) daisies in a field?
   b) butterflies?
   c) woodlice?
3. Five daisy plants are found in a 0.25 m² quadrat. How many would there be in a 100 m² field?
4. Why is it best to sample several areas in the field and take an average?

Stretch Yourself

1. A total of 30 snails are collected in an area, marked and released. When another sample is captured there are 35 snails and 5 are marked. What is an estimate of the snail population?
2. Different animals live on different parts of a rocky shore on the way down to the sea. Referring to tides, explain why the animals show zonation.

Photosynthesis

The Reactions of Photosynthesis

Plants make their own food by a process called **photosynthesis**. They take in carbon dioxide and water and turn them into sugars, releasing oxygen as a waste product. The process needs the energy from sunlight and this is trapped by the green pigment **chlorophyll**.

$$6CO_2 + 6H_2O \rightarrow C_6H_{12}O_6 + 6O_2$$

Boost Your Memory

Try this for remembering the equation for photosynthesis: **C**ertain **w**orms **e**at **g**rass **o**utside (**c**arbon dioxide, **w**ater, **e**nergy, **g**lucose, **o**xygen).

Where does it happen?

Photosynthesis occurs mainly in the leaves.

The leaves are specially adapted for photosynthesis in a number of ways:

- A broad shape – provides a large surface area to absorb light and CO_2.
- A flat shape – the gases do not have too far to diffuse.
- Contain a network of veins – supply water from the roots and take away the products.
- Contain many chloroplasts in the palisade layer near the top – this traps the maximum amount of light.
- Pores called stomata (singular stoma) and air spaces – they allow gases to diffuse into the leaf and reach the cells.

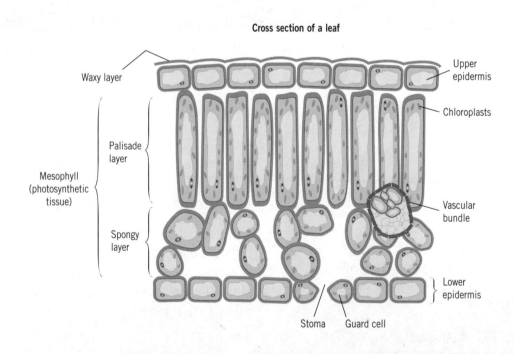

Cross section of a leaf

Photosynthesis Experiments

The understanding of the process of photosynthesis has changed considerably over time:
- Greek scientists thought that plants gained mass only by taking in minerals from the soil.
- Van Helmont in the 17th century worked out that plant growth could not be solely due to minerals from the soil. He found that the mass gained by a plant was more than the mass lost by the soil.
- In the 18th century Priestley showed that oxygen is produced by plants.

More modern experiments using isotopes have increased our understanding of photosynthesis. Isotopes of carbon can be used that behave in the same way chemically as carbon but which can be followed in the reactions because they are radioactive.

These experiments have shown that photosynthesis is a two-stage process:
- Light energy is used to split water, releasing oxygen gas and hydrogen atoms.
- Carbon dioxide gas combines with the hydrogen to make glucose and water.

Build Your Understanding

The rate of photosynthesis can be increased by providing:
- More light.
- More carbon dioxide.
- An optimum temperature.

Any of these factors can be limiting factors.

A limiting factor is something that controls how fast a reaction occurs. If more light is provided, it increases photosynthesis because more energy is available. After a certain point something else limits the rate.

Similarly more carbon dioxide increases the rate up to a point because more raw materials are present. Increasing the temperature makes enzymes work faster, but high temperatures denature enzymes.

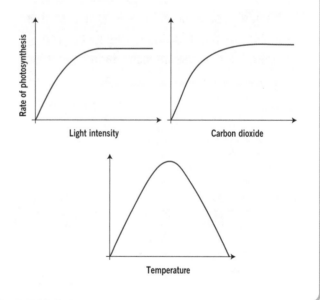

✓ Maximise Your Marks

Many candidates think that plants respire at night and photosynthesise during the day. To get an A* you must realise that plants carry out respiration all the time. Fortunately for us, during the day, they photosynthesise much faster than they respire, so overall release more oxygen than they take in.

? Test Yourself

1. What is the job of chlorophyll in photosynthesis?
2. Which cells in the leaf have most chloroplasts?
3. What are stomata?
4. Why do leaves have veins?

★ Stretch Yourself

1. In the reactions of photosynthesis, is oxygen released from water, carbon dioxide or both?
2. Why do high temperatures stop photosynthesis happening?

Food Production

Plants Need Minerals

Once plants have made sugars such as glucose by photosynthesis, they can convert it into many different things they need in order to grow:

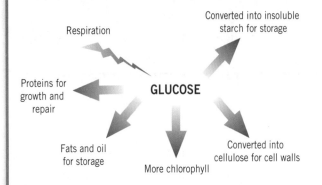

To produce these chemicals, plants need various minerals from the soil:

- **Nitrates** as a supply of nitrogen to make amino acids and proteins.
- **Phosphates** to supply phosphorus to make DNA and cell membranes.
- **Potassium** to help enzymes in respiration and photosynthesis.
- **Magnesium** to make chlorophyll.

Without these minerals plants do not grow properly. Farmers must therefore make sure that they are available in the soil.

Intensive Food Production

The human population is increasing and so there is a greater demand for food. This means that many farmers now use **intensive farming** methods.

Intensive farming means trying to obtain as much food as possible from the land. There are a number of different food production systems that use intensive methods:

Food production systems that use intensive methods

Fish farming

Fish are kept in enclosures away from predators. Their food supply and pests are controlled.

Glasshouses

Plants can be grown in areas where the climate would not be suitable. They can also produce crops at different times of the year.

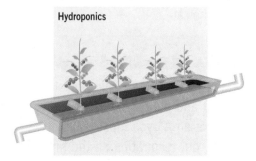

Hydroponics

Plants are grown without soil. They need extra support but their mineral supply and pests are controlled.

Build Your Understanding

Farmers use a number of intensive farming techniques to help increase their yield but it is argued that the damage caused by some of these techniques does not justify the increase in food production:

- They use pesticides to kill pests that might eat the crop.
- They use herbicides to kill weeds that would compete with the crop.
- They keep animals indoors so that they do not waste energy keeping warm or moving about.
- They provide the plants with chemical fertilisers for growth.

Organic Food Production

Many people think that intensive farming is harmful to the environment and cruel to animals. Farming that does not use the intensive methods is called **organic farming**. Organic farming uses a number of different techniques:

Technique	Details
Use of manure and compost	These are used instead of chemical fertilisers and provide minerals for the plant
Crop rotation	Farmers do not grow the same crop in the same field year after year; this stops the build-up of pests and can reduce nutrient depletion of the soil
Use of nitrogen-fixing crops	These crops contain bacteria that add minerals to the soil
Weeding	This means that chemical herbicides are not needed
Varying planting times	This can help to avoid times that pests are active
Using biological control	Farmers can use living organisms to help to control pests; the organisms may eat the pests or cause disease

Preserving Food

Preservation method	How it is done	How it works
Canning	Food is heated in a can and the can is sealed	The high temperature kills the microorganisms, and oxygen cannot get into the can after it is sealed
Cooling	Food is kept in a refrigerator at about 5°C	The growth and respiration of the decomposers slow down at low temperature
Freezing	Food is kept in a freezer at about −18°C	The decomposers cannot respire or reproduce
Drying	Dry air is passed over the food	Microorganisms cannot respire or reproduce without water
Adding salt or sugar	Food is soaked in a sugar solution or packed in salt	The sugar or salt draws water out of the decomposers
Adding vinegar	The food is soaked in vinegar	The vinegar is too acidic for the decomposers

Although gardeners want decay to happen in their compost heaps, people do not want their food to decay before they can eat it. **Food preservation** methods reduce the rate of decay of foods. There are many ways to preserve food. Most stop decay by taking away one of the factors that decomposers need.

? Test Yourself

1. Why do plants need nitrates?
2. Why does a plant look yellow if grown with a lack of magnesium?
3. What is hydroponics?
4. Why does food still go bad in a refrigerator?

★ Stretch Yourself

1. Suggest one problem with using large quantities of chemical pesticides to kill insect pests.
2. In intensive farming, why is the food brought to pigs rather than letting them find food?

Transport in Animals

Blood

Blood is made up of a liquid called **plasma**.

Plasma carries chemicals such as dissolved food, hormones, antibodies and waste products around the body.

Cells are also carried in the plasma. They are adapted for different jobs.

Red blood cells are shaped like a biconcave disc. They contain haemoglobin which can carry oxygen around the body.

The haemoglobin in red blood cells reacts with oxygen in the lungs, forming oxyhaemoglobin. In the tissues, the reverse of this reaction happens and oxygen is released:

haemoglobin + oxygen ⇌ oxyhaemoglobin

White blood cells can change shape to engulf and destroy disease causing organisms. They can also produce antibodies.

Platelets are responsible for clotting the blood.

A red blood cell

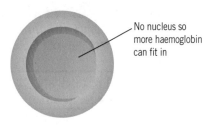
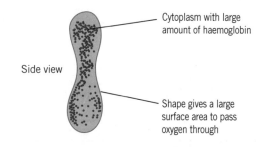

Side view

✓ Maximise Your Marks

The shape of red blood cells does not allow them to carry more oxygen, but the increased surface area to volume ratio lets them gain or lose it more quickly

Blood Vessels

The blood is carried around the body in **arteries**, **veins** and **capillaries**.

Arteries	Veins	Capillaries
Carry blood away from the heart	Carry blood back to the heart	Join arteries to veins
Have thick, muscular walls because the blood is under high pressure	Have valves and a wide lumen because the blood is under low pressure	Have permeable walls so that substances can pass in to and out of the tissues

💡 Boost Your Memory

You need to remember that arteries carry blood **a**way from the heart and veins carry it back **in**to the heart.

The Heart

The heart is made up of four chambers.

The top two chambers are called **atria** and they receive blood from veins.

The bottom two chambers are **ventricles**. They pump the blood out into arteries.

The top two chambers, the atria, fill up with blood returning in the **vena cavae** and **pulmonary veins**. The two atria then contract together and pump the blood down into the ventricles. The two ventricles then contract, pumping blood out into the **aorta** and **pulmonary arteries** at high pressure.

In the heart are two sets of valves, whose function is to prevent blood flowing backwards.

In between the atria and the ventricles are the **bicuspid** and **tricuspid valves**.

These valves stop blood flowing back into the atria when the ventricles contract. The pressure of blood closes the flaps of the valves and the tendons stop the flaps turning inside out.

There are also **semi-lunar** valves between the ventricles and the arteries.

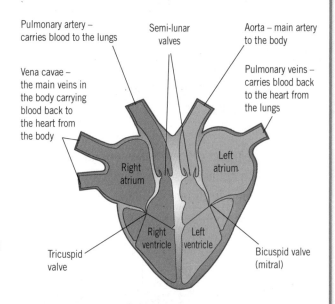

Cross section of a heart

✓ Maximise Your Marks

Make sure that you can spot that the muscle wall of the left ventricle is always thicker than that of the right ventricle. This is because it has to pump blood all round the body compared with the short distance to the lungs.

Build Your Understanding

Mammals have a **double circulation**.

This means that the blood has to pass through the heart twice on each circuit of the body.

Deoxygenated blood is pumped to the lungs and the oxygenated blood returns to the heart to be pumped to the body.

The advantage of this system is that the pressure of the blood stays quite high and so it can flow faster around the body.

Because of the double circulation the heart is really two pumps in one:
- The right side pumps the blood to the lungs.
- The left side pumps it to the rest of the body.

❓ Test Yourself

1. What is the job of platelets?
2. Why do red blood cells lack a nucleus?
3. Why do veins have valves?
4. What blood vessel carries blood from the heart to the lungs?

★ Stretch Yourself

1. Why is the right side of the heart coloured blue in the diagram?
2. Some people have a defect in the bicuspid valve. Explain why this can lead to a build up of blood in the blood vessels of the lungs.

Transport in Plants

Xylem and Phloem

Plants have two different tissues that are used to transport substances. They are called **xylem** and **phloem**.

Xylem vessels and phloem tubes are gathered together into collections called **vascular bundles**. They are found in different regions of the leaf, stem and root.

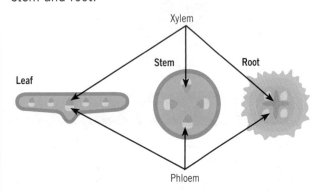

Xylem vessels and phloem tubes are different in structure and do different jobs:

Xylem	Phloem
Carries water and minerals from roots to the leaves	Carries dissolved food substances both up and down the plant
The movement of water up the plant and out of the leaves is called transpiration	The movement of the dissolved food is called translocation
Made of vessels which are hollow tubes made of thickened dead cells	Made of columns of living cells

💡 Boost Your Memory

Remember: phloem for food and xylem for water.

The Movement of Water

Water enters the plant through the **root hairs** by **osmosis**.

The root hair cells increase the surface area for the absorption of water.

Water then passes from cell to cell by osmosis until it reaches the centre of the root.

The water enters xylem vessels in the root and then travels up the stem.

Water enters the leaves and evaporates.

It then passes through the **stomata** by **diffusion**.

This loss of water is called **transpiration** and it helps to pull water up the xylem vessels.

Various environmental conditions can affect the transpiration rate.

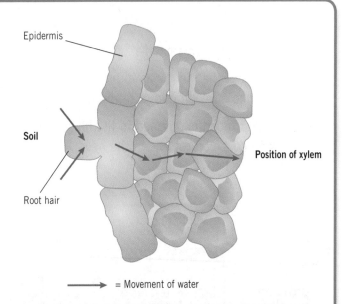

✓ Maximise Your Marks

Remember that it is osmosis that brings the water into the leaf and into the xylem, but water does not move up the xylem by osmosis. It is 'sucked up' by evaporation from the leaves.

Transpiration Rate

The rate of transpiration depends on a number of factors:
- **Temperature** – warm weather increases the kinetic energy of the water molecules so they move out of the leaf faster.
- **Humidity** – damp air reduces the concentration gradient so the water molecules leave the leaf more slowly.
- **Wind** – the wind blows away the water molecules so that a large diffusion gradient is maintained.
- **Light** – light causes the stomata to open and so more water is lost.

The factors that speed up transpiration will also increase the rate of water uptake from the soil. If water is scarce, or the plant roots are damaged, the plants chances of survival is increased if transpiration can be slowed down. Plants do this by wilting, or can be cut so that they can grow new roots.

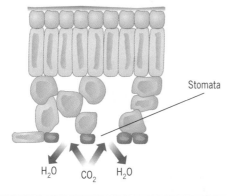

Stomata

H_2O CO_2 H_2O

✓ Maximise Your Marks

If a question asks you to 'give one factor that increases transpiration rate', make sure that you write 'an increase in temperature', not just 'temperature'. Many candidates lose marks in this way.

Build Your Understanding

When plants are short of water, they do not want to waste it through transpiration. The trouble is they need to let carbon dioxide in, so water will always be able to get out. Water loss is kept as low as possible in several ways:
- Photosynthesis only occurs during the day, so the stomata close at night to reduce water loss. The guard cells lose water by osmosis and become flaccid. This closes the pores.
- The stomata are on the underside of the leaf. This reduces water loss because they are away from direct sunlight and protected from the wind.
- The top surface of the leaf, facing the Sun, is often covered with a protective waxy layer.

Although transpiration is kept as low as possible, it does help plants by cooling them down and supplying leaves with minerals. It also provides water for support and photosynthesis.

❓ Test Yourself

1. In which direction in a plant stem do water and minerals move?
2. What is translocation?
3. Where in a plant root is xylem found?
4. What is the function of root hair cells?

★ Stretch Yourself

1. Why is it impossible for plants to prevent all water loss from the leaves?
2. What causes stomata to close when a plant wilts?

Digestion and Absorption

Digestion

The job of the digestive system is to break down large food molecules into small soluble molecules. This is called **digestion**.

Digestion happens in two main ways – **physical** and **chemical** digestion.

Physical digestion occurs in the mouth where the teeth break up the food into smaller pieces.

Chemical digestion is caused by digestive enzymes that are released at various points along the digestive system. Most enzymes work inside cells, controlling reactions. Some enzymes pass out of cells and work in the digestive system. These enzymes digest our food, making the molecules small enough to be absorbed.

The food is moved along the gut by contractions of the muscle in the lining of the intestine. This process is called **peristalsis**.

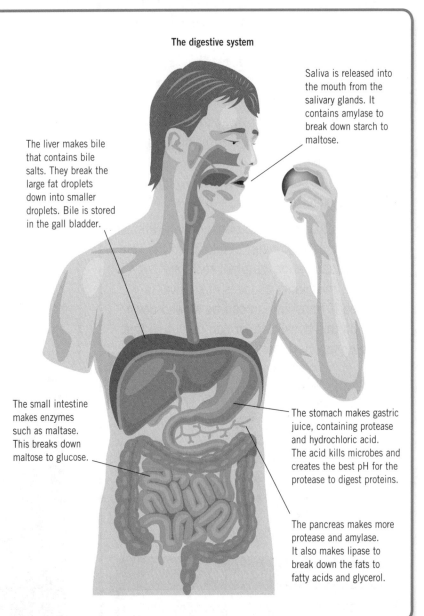

The digestive system

Saliva is released into the mouth from the salivary glands. It contains amylase to break down starch to maltose.

The liver makes bile that contains bile salts. They break the large fat droplets down into smaller droplets. Bile is stored in the gall bladder.

The small intestine makes enzymes such as maltase. This breaks down maltose to glucose.

The stomach makes gastric juice, containing protease and hydrochloric acid. The acid kills microbes and creates the best pH for the protease to digest proteins.

The pancreas makes more protease and amylase. It also makes lipase to break down the fats to fatty acids and glycerol.

Build Your Understanding

To make the digestive enzymes work at an optimum rate, the digestive system provides the best conditions:
- Each enzyme has a different optimum pH. Protease in the stomach works best at about pH 2, but a different protease made by the pancreas works best at about pH 9.
- Physical digestion helps to break the food into smaller particles, thereby increasing the surface area of the food particles. Bile salts **emulsify** fat droplets, breaking them into smaller droplets so lipase can work faster.

✓ Maximise Your Marks

Be careful not to say that bile salts break down fats. Make sure that you say 'into fat droplets', otherwise it sound like bile salts are doing the same job as lipase.

Absorption

In the small intestine, small digested food molecules are absorbed into the bloodstream by diffusion. The inside of the small intestine is permeable and has a large surface area over which absorption can take place.

The lining of the small intestine contains two types of vessel that absorb the products of digestion:
- **Capillaries** absorb food and take it to the liver via the **hepatic portal vein**.
- **Lacteals** absorb mainly the products of fat digestion and empty them into the bloodstream.

Build Your Understanding

A number of factors increase the surface area of the small intestine and so speed up the rate of absorption:
- The human small intestine is over 5 metres long.
- The inner lining is folded.
- The folds are covered with finger-like projections called villi.
- The villi are further covered by smaller projections called microvilli.

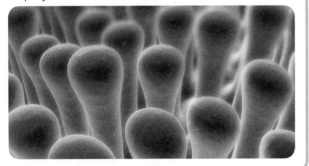

Other Uses of Digestive Enzymes

Microorganisms also make digestive enzymes. Decay organisms such as certain bacteria and fungi release these enzymes on to the food and take up the soluble products. These organisms are called **saprophytes**.

Scientists have used microorganisms such as saprophytes to supply enzymes for a number of uses.

- **Proteases** and **lipases** are used in biological washing powders.
- **Proteases** are used to pre-digest protein in some baby foods.
- **Amylases** are used to convert starch into sugar syrup.
- **Isomerase** is used to convert glucose into fructose which is sweeter.

❓ Test Yourself

1. Why does bread start to taste sweet if it is chewed for several minutes?
2. What is the function of the gall bladder?
3. What are the products of fat digestion?
4. Why is fructose used in sweets rather than glucose?

⭐ Stretch Yourself

1. The lining of the stomach is protected by mucus. Why does it need to be protected?
2. People who have coeliac disease may have many of their villi destroyed. What effect might this have on the process of absorption? Explain your answer.

Practice Questions

 Complete these exam-style questions to test your understanding. Check your answers on page 124. You may wish to answer these questions on a separate piece of paper.

1 The following structures are found in plant and animal cells. Match words **A**, **B**, **C**, and **D**, with numbers **1–4** in the sentences. (4)

A mitochondria **B** cell wall **C** vacuole **D** cell membrane

All organisms release energy from food. This largely happens in the _____1_____. Cells take up water by osmosis because the _____2_____ is partially permeable. The _____3_____ stores some sugars and salts. Plant cells are limited to how much water they can take up because the _____4_____ resists the uptake of too much water.

2 Arthur wants to measure how fast a plant photosynthesises at different light intensities. The diagram shows the apparatus he uses.

Arthur includes the following steps in his method:
- He uses the same piece of pondweed for the complete investigation.
- He adds sodium hydrogen carbonate to the water to provide the plant with carbon dioxide.
- He times five minutes using the minute hand of his watch and counts the number of bubbles given off.
- He counts the bubbles three different times at each different light intensity.
- He repeats this with the light at different distances from the pondweed.

a) Why does Arthur choose pondweed for his experiment? (2)

b) What is the main gas found in the bubbles? (1)

c) Write down the step that helps to make Arthur's experiment valid. (1)

d) Suggest one way that Arthur could make his experiment more accurate. (1)

3 All the cells in the human body have about 20 000 genes. Scientists have studied some organs to see how many of these genes are used by cells in each organ. This number is shown below.

- Liver 2091
- Kidney 712
- Heart 1195
- Pancreas 1094
- Small intestine 297

a) Write down precisely where in a cell the genes are found. (2)

b) Genes are code for the production of proteins. Explain how each gene can code for a different protein. (2)

c) What percentage of its genes does each pancreas cell actually use? (1)

d) Which organ in the list would you expect to carry out the most chemical reactions? Explain your answer. (2)

4 Complete these sentences by writing the correct words in the gaps. (5)

Starch molecules are too large to be able to pass into the bloodstream and so need to be _____ first. This digestion begins in the _____. An enzyme called _____ breaks down starch into maltose. Maltose is then digested into _____ in the small intestine. Absorption then occurs and this is speeded up by the presence of tiny projections on the wall of the small intestine called _____.

5 The diagrams show three different blood vessels.

A _____ B _____ C _____

Write the name of each type of blood vessel under the correct diagram. (3)

How well did you do?

0–8 Try again | 9–14 Getting there | 15–19 Good work | 20–24 Excellent!

Distance, Speed and Velocity

Distance, Speed and Velocity

Speed is measured in metres per second (m/s) or kilometres per hour (km/h):
- An athlete running with a speed of 5 m/s travels a distance of 5 metres in one second and 10 metres in two seconds.
- An athlete with a faster speed of 8 m/s travels further, 8 metres, in each second and takes less time to complete his journey.

To calculate speed:

$$\text{speed (m/s)} = \frac{\text{distance (m)}}{\text{time (s)}}$$

💡 Boost Your Memory

Some students find using the following triangle method useful to rearrange a formula.

Example: $\text{speed} = \dfrac{\text{distance}}{\text{time}}$

To calculate distance or time:
- Write the formula into a triangle, so that distance is 'over' time. This means putting distance at the top, time can go in either of the other corners.

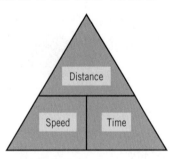

- To find the **distance**, cover the word distance with your finger and look at the position of speed and time. They are side by side, so **distance = speed × time**.
- To find the **time**, cover the word time with your finger and distance is 'over' speed, so

$$\text{time} = \frac{\text{distance}}{\text{speed}}$$

Build Your Understanding

There are two ways of looking at a journey:
- The distance travelled can only increase, or stay the same, so speed is always a positive number.
- The direction travelled is important, so that travel in one direction is a positive distance and in the opposite direction is a negative distance. Sometimes, distance in a given direction is called displacement.

Quantities that have magnitude and direction are called vectors.

Velocity is a vector, because velocity is speed in a given direction. For example, a dog walks in a positive direction and then back again with a constant speed of 2 m/s, so he walks with a velocity of +2 m/s and then with a velocity of –2 m/s.

$$\text{velocity (m/s)} = \frac{\text{displacement (m)}}{\text{time (s)}}$$

✓ Maximise Your Marks

When you are doing calculations:
- Always show your working – if you make a mistake in the calculation you may still get some marks.
- Do not write down *only* the triangle. You will not get marks for using the correct formula or equation.

Distance-Time Graphs

On a **distance-time graph**:
- A horizontal line means the object is stopped.
- A straight line sloping upwards means it has a steady speed.

The steepness, or **gradient**, of the line shows the speed:
- A steeper gradient means a higher speed.
- A curved line means the speed is changing.

Distance-Time Graphs (cont.)

Example: Distance-time graph for a cycle ride.

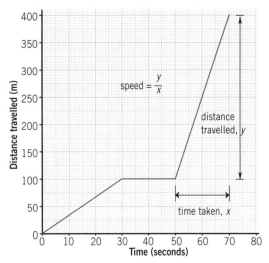

Between 30 s and 50 s the cyclist stopped. The graph has a steeper gradient between 50 s and 70 s than between 0 s and 30 s. The cyclist was travelling at a greater speed.

To calculate a speed from a graph, work out the gradient of the straight line section.

speed = $\frac{y}{x}$ where y = 400 m − 100 m = 300 m

and x = 70 s − 50 s = 20 s.

speed = $\frac{300 \text{ m}}{20 \text{ s}}$ = 15 m/s.

Build Your Understanding

If the direction of travel is being considered:
- A negative displacement is in the opposite direction to a positive displacement.
- A straight line sloping upwards or downwards means steady speed.
- Upwards means a steady positive velocity, and downwards means a steady negative velocity.

Example: Displacement-time graph for a journey from home.

A boy starts from home (0 km) and walks to a shop, home again and then in the opposite direction to the shop.

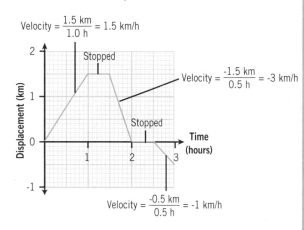

Average Speed and Instantaneous Speed

The **average speed** of the cyclist for the total journey shown on the graph is:

= $\frac{\text{total distance}}{\text{total time}}$ = $\frac{400 \text{ m}}{70 \text{ s}}$ = 7.62 m/s

This is not the same as the **instantaneous speed** at any moment because the speed changes during the journey. If you calculate the average speed over a shorter time interval you get closer to the instantaneous speed.

? Test Yourself

1. What is the difference between instantaneous speed and average speed?
2. In the graph of the cycle ride, what is the speed during the first 30 seconds?
3. A car travels 288 km in three hours. Calculate the speed in km/h.
4. A car travels at a speed of 12 m/s. How long will it take to travel 1.44 km?

★ Stretch Yourself

1. In the graph of the journey from home:
   a) What was the displacement after the first 36 minutes?
   b) When was the speed greatest?
   c) How can you tell this without doing any calculations?

Speed, Velocity and Acceleration

Acceleration

A change of velocity is called **acceleration**. Speeding up, slowing down and changing direction are all examples of acceleration.

Acceleration is the change in velocity per second. It is measured in metres per second squared (m/s²).

Example: If a car accelerates from 0 to 27 m/s (about 60 mph) in six seconds the change in velocity is 27 m/s, the acceleration:

$$= \frac{27 \text{ m/s}}{6 \text{ s}} = 4.5 \text{ m/s}^2$$

Calculating Acceleration

$$\text{acceleration (m/s}^2\text{)} = \frac{\text{change in velocity (m/s)}}{\text{time taken (s)}}$$

$$a = \frac{(v-u)}{t}$$

Where a is the acceleration of an object whose velocity changes from initial velocity u to final velocity, v in time t.

Example: A car accelerates from 14 m/s to 30 m/s in 8 s. The acceleration:

$$a = \frac{(30 \text{ m/s} - 14 \text{ m/s})}{8 \text{ s}} = \frac{16 \text{ m/s}}{8 \text{ s}} = 2 \text{ m/s}^2$$

Speed-Time Graphs

Plotting the speed of an object against the time gives a graph like this:

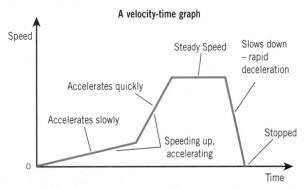

- A **positive slope (gradient)** means that the speed is increasing – the object is accelerating.
- A **horizontal line** means that the object is travelling at a steady speed.
- A **negative slope (gradient)** means the speed is decreasing – negative acceleration.
- A **curved slope** means that the acceleration is changing – the object has **non-uniform acceleration**.

Tachographs are instruments that are put in lorry cabs to check that the lorry has not exceeded the speed limit and that the driver has stopped for breaks. They draw a graph of the speed against time for the lorry.

✓ Maximise Your Marks

Always check carefully whether a graph is a speed-time graph or a distance-time graph.

Build Your Understanding

On true **speed-time graphs** the speed has only positive values. On **velocity-time graphs** the velocity can be negative.

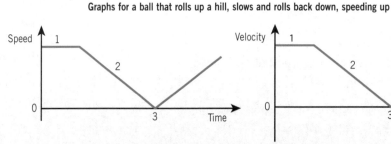

Graphs for a ball that rolls up a hill, slows and rolls back down, speeding up

Acceleration from a Graph

The acceleration is the gradient of a velocity-time graph.

Example: A graph of velocity against time for a car journey.

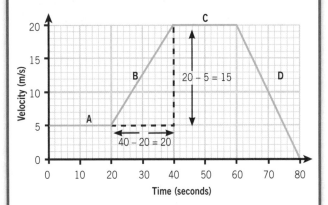

In part B of the graph the car accelerates from 5 m/s to 20 m/s in (40 − 20)s = 20 seconds. The gradient:

$$= \frac{y}{x} = \frac{15}{20} = 0.75$$

So the acceleration = 0.75 m/s^2.

💡 Boost Your Memory

Remember the gradient is: $\frac{\text{rise}}{\text{run}}$

Build Your Understanding

On a velocity-time graph the area between the graph and the time axis represents the distance travelled.

Example: Distance travelled on a bike journey.

The distance travelled: **= area under A + area under B**

Area A = area of pink triangle = ½ (10 m/s × 20 s) = 100 m

Area B = area of blue rectangle = 10 m/s × (60 − 20)s = 400 m

Distance travelled = 100 m + 400 m = 500 m

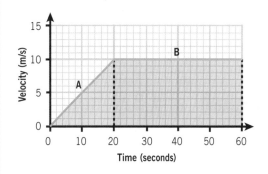

✓ Maximise Your Marks

Take care with units. You may need to change minutes or hours to seconds, or kilometres to metres:

$$36 \text{ km/h} = \frac{36 \times 1000}{60 \times 60} \text{ m/s} = 10 \text{ m/s}$$

Also remember to write down the units for your answer.

❓ Test Yourself

1. A car accelerates from 0 to 24 m/s in 8 s. What is the acceleration?
2. What does a horizontal line on a speed-time graph tell you?
3. From the graph of the car journey, what is the acceleration:
   a) in part C
   b) in part D?

⭐ Stretch Yourself

1. Sketch a velocity time graph for a car that speeds up in a negative direction, travels at a steady speed, and then slows down and stops.
2. From the graph of the car journey, what is the distance travelled in:
   a) in part C
   b) in part D?

Forces

Mass and Weight

Mass is measured in **kilograms** (kg). It is the amount of matter in an object. An object has the same mass everywhere, on the Earth, the Moon, or in outer space.

Weight is a force and is measured in **newtons** (N). Weight is the force of gravity attracting the mass towards the centre of the Earth:

weight (N) = mass (kg) × gravitational field strength (N/kg)

Build Your Understanding

In outer space there is no gravity so all objects are weightless. The Moon's gravitational attraction is only one sixth of the Earth's, so objects on the Moon have only one sixth their weight on Earth, though their mass remains constant.

💡 Boost Your Memory

To remember the differences between mass and weight think of a tin of beans:
- It's weightless in outer space.
- It has less weight on the Moon.
- Its mass changes if you eat the beans.

Resultant and Balanced Forces

Forces have size and direction. The length of the arrow on a diagram represents the size of the force.

When several forces act on an object the effect is the same as one force in a certain direction. This is called the **resultant force**. Forces combine to give a resultant force. If the resultant force is zero the forces on the object are **balanced**.

A resultant force changes the velocity of an object. This idea is known as **Newton's First Law of Motion**:

If the resultant force on an object is zero, the object will remain stationary or continue to move at a steady speed in the same direction.

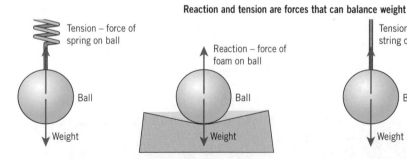

Resultant forces

More Balanced Forces

When forces on an object are balanced it does not fall.

Reaction and tension are forces that can balance weight

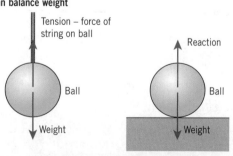

The upward force of the spring, or the foam, trying to return to their original shape balances the weight.

Even though the changes in shape are too small for us to see, the restoring force – the tension in the string or the reaction from the floor, balances the weight.

Resistance to Motion-Friction Forces

When one object slides, or tries to slide, over another, there is **friction**, the resistive force between the two surfaces.

Air resistance is the resistive force that acts against objects moving through the air.

Drag is the resistive force on objects moving through liquids or gases. Drag is larger in liquids.

These resistive forces:
- Always act against the direction of motion.
- Are zero when there is no movement.
- Increase as the speed of the object increases.

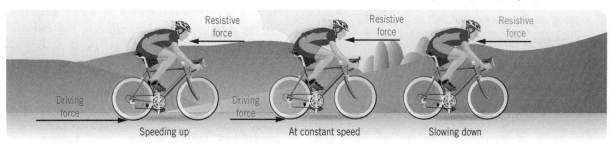

Forces when Falling

The resultant force on a skydiver changes.

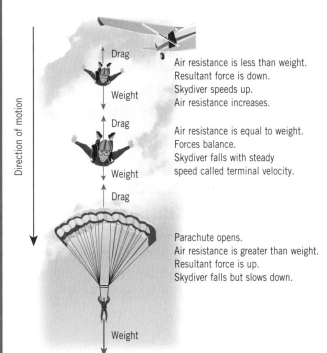

Air resistance is less than weight.
Resultant force is down.
Skydiver speeds up.
Air resistance increases.

Air resistance is equal to weight.
Forces balance.
Skydiver falls with steady speed called terminal velocity.

Parachute opens.
Air resistance is greater than weight.
Resultant force is up.
Skydiver falls but slows down.

Falling objects reach a steady speed called the terminal velocity, when: **drag = weight**.

In a vacuum, there is no air resistance so objects continue to fall with acceleration due to gravity of 10 m/s^2.

A falling object eventually reaches a terminal velocity

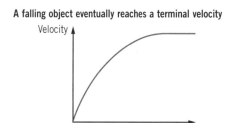

✓ Maximise Your Marks

Remember if an object is travelling at a steady speed there is no resultant force on it. When an object is slowing down the resultant force on it is opposite to its direction.

It's a common mistake to think that there is always a force in the direction of movement.

❓ Test Yourself

1. What is the weight of 2 kg of sugar?
2. The driving force is 3000 N and friction is 900 N. What is the resultant force?
3. A car travels with steady speed of 100 km/h in a straight line. What is the resultant force?

⭐ Stretch Yourself

1. How much does a 100 g apple weigh a) on Earth and b) on the Moon?
2. The driving force is 2500 N, friction is 800 N and air resistance is 1700 N. What is the resultant force?

Acceleration and Momentum

Force and Acceleration

A resultant force on an object causes it to accelerate. The acceleration is:
- Larger for a larger force.
- Smaller for a larger mass.
- In the same direction as the force.

For a resultant force on an object:
force (N) = mass (kg) × acceleration (m/s^2)
F = ma
Where F = force, m = mass and a = acceleration.

Force and Momentum

The **momentum** of an object is:
- Larger for a larger velocity.
- Larger for a larger mass.
- In the same direction as the velocity.
- Measured in units called kg m/s or Ns (they are the same).

momentum (kgm/s or Ns) = mass (kg) × velocity (m/s)

where mv = momentum, v = velocity.

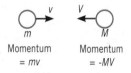

When a resultant force acts on an object, it causes a change in momentum in the same direction as the force (because the velocity changes).

The bigger the force and the longer the time that the force acts, the bigger the change in momentum:

change in momentum (kg m/s or Ns) = force (N) × time (s)

💡 Boost Your Memory

To remember what momentum depends on, think about 10 pin bowling. You are more likely to knock pins down with more momentum. This could be a ball with a lot of mass or a ball with a high velocity.

Build Your Understanding

Newton's Second Law of Motion states:

When a resultant force acts on an object, it causes a change in momentum in the same direction as the force. The resultant force equals the rate of change of momentum.

$$\text{force (N)} = \frac{\text{change in momentum}}{\text{time (s)}} \text{ (kgm/s or Ns)}$$

$$F = \frac{(mv - mu)}{t}$$

Where t = time, u = initial velocity and v = final velocity.

The equation $F = ma$ is another way of saying this because acceleration:

$$a = \frac{(v-u)}{t}$$

$$F = \frac{m(v-u)}{t} = \frac{(mv-mu)}{t}$$

You do not need to be able to show or explain this.

Safer Collisions

In a collision, a force brings your body to a sudden stop. The larger the stopping force on the body the more it is damaged. To reduce damage we must reduce the force:

change of momentum = force × time

For the same change in momentum, to reduce the force we must increase the time to stop.

If the collision takes place over a longer time, say 0.5 s instead of 0.05 s – 10 times as long – then the stopping force will only be one tenth of the size.

Stopping Safely

These safety features work by increasing the time taken for collision:

- **Crumple zones** in the car. The front and back of the car are designed to crumple in a collision, increasing the distance and time over which the occupants are brought to a stop.

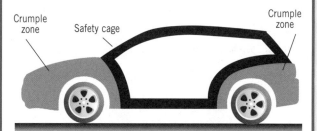

- The body hits the **airbag**, which is compressed, increasing the distance the body moves and the time it takes to stop.
- **Seatbelts** are designed to stretch slightly so that the body moves forward and comes to a stop more slowly with a smaller deceleration than it would if it hit the windscreen or front seats. After a collision the seatbelts should be replaced because they are not elastic and once stretched will not return to their shape and may break in a second collision.
- **Cycle and motorcycle helmets** contain a layer of material which will compress on impact so that the skull is brought to a stop more slowly. They should be replaced after a collision as the material will be damaged and may not protect you again.

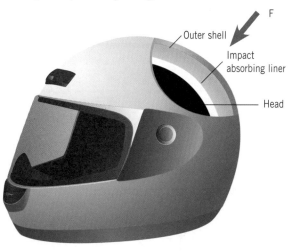

Build Your Understanding

Other examples of reducing the force by increasing the time taken to stop include:
- Crash barriers that crumple on impact.
- Bending your knees when you land after jumping.
- Bubble wrap.
- Sprung floors in gyms.

✓ Maximise Your Marks

Do not confuse momentum with energy. Momentum has a direction and is measured in Ns or kg m/s. Energy, like mass, has no direction and it is measured in Joules (J). 1 J = 1 Nm – a different unit.

❓ Test Yourself

1. Calculate the resultant force on a 1200 kg car accelerating at 3 m/s^2.
2. Calculate the momentum of a ball of mass 2 kg and velocity 5 m/s.
3. Calculate the momentum of a ball of mass 200 g and velocity 8 m/s travelling in the opposite direction.
4. A force of 50 N acts on a stationary object for 12 seconds. Calculate its gain in momentum.

⭐ Stretch Yourself

1. A runaway truck mass 1000 kg and velocity 12 m/s came to a sudden stop in 0.002 s.
   a) Calculate the stopping force on the truck.
   b) A crash barrier would have stopped it over 0.5 s. What would the stopping force have been?
   c) What difference would this make?

Pairs of Forces: Action and Reaction

Interaction Pairs

When two objects interact there is always an **interaction pair of forces**. As these skaters show, the boy cannot push the girl without the girl pushing the boy.

In an interaction pair of forces, the two forces:
- Are always equal in size and opposite in direction.
- Always act on different objects.
- Are always the same type of force (for example, contact forces, gravitational forces, or magnetic forces).

This idea is known as **Newton's Third Law of Motion** which states:

When two objects interact, the forces they exert on each other are equal and opposite and are called action and reaction forces.

Friction and Getting Started

Friction is the reaction force needed for walking or wheeled transport.

- **Action force** is where the wheel pushes back on the road.
- **Reaction force** is where the road pushes forward on the wheel, which sends the wheel forward.

> ### Boost Your Memory
>
> To remember how friction gets you moving, think of trying to cycle on a frictionless icy surface. There is no reaction force. You would slip backwards and never move forward.

When you walk your foot pushes back on the ground and the ground pushes your foot forward.

Forces that make a wheel move forward
Action Force   Reaction Force

Build Your Understanding

The weight of an object is the gravitational attraction towards the centre of the Earth. The other force of this interaction pair acts on the Earth. It is the gravitational attraction of the object attracting the whole Earth.

We do not notice this effect because the mass of the Earth is so large.

An action and reaction pair of forces

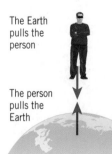

The Earth pulls the person

The person pulls the Earth

Newton's third law: pairs of forces that are equal in size and opposite in direction.

Different Forces

Do not confuse interaction pairs of forces, which act on different objects, with balanced forces, which act on the same object.

The reaction force from a surface *balances* the weight of the object. It is not the reaction pair of the weight, because:
- It is a different type of force (a contact force, not gravitational).
- It acts on the same object.

Rockets and Jet Engines

Rockets and jet engines both produce hot exhaust gases. These are pushed out of the back of the engines. There is an equal and opposite reaction force that sends the rocket or jet forward.

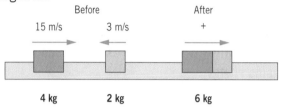

Exhaust gases gain momentum in this direction

Rocket gains momentum in this direction

Recoil

When a bullet leaves a gun, action and reaction, or **conservation of momentum**, tell us the gun must recoil.

Example: A 0.8 g paintball is fired at 80 m/s from a 3 kg paintball marker.

m = 0.8 g, v = 80 m/s paintball

M = 3 kg, $-V$? paintball marker

Conservation of momentum: $mv = MV$

$$V = \frac{0.8 \text{ g} \times 80 \text{ m/s}}{3000 \text{ g}} = 0.021 \text{ m/s}$$

Build Your Understanding

When two objects, collide or explode apart there is an equal and opposite force on each object for the same time, so the change in the momentum of the objects is equal and opposite. Momentum is conserved. The total momentum of two objects before collision or explosion is the same as the total momentum after.

Example: When two objects collide and stick together.

Before: 15 m/s (4 kg), 3 m/s (2 kg)
After: + (6 kg)

Before the collision momentum:
= 4 kg × 15 m/s + 2 kg × (–3) m/s
= (60 – 6) kg m/s = 54 kg m/s

After the collision: momentum = 6 kg × v

So, because of conservation of momentum:

$v = \dfrac{54}{6} = 9$ m/s

v = 9 m/s

? Test Yourself

1. A girl pushes on a wall with a force of 5 N. Describe the reaction force.
2. Why is it difficult to walk on ice?
3. How does a rocket move in outer space where there is nothing to push against to get moving?
4. When you release a partly inflated balloon it flies around as it deflates. Explain why.

★ Stretch Yourself

1. A book is placed on a table. What are the two interaction pairs of forces?
2. A toy car with mass 0.5 kg and speed 4 m/s collides with a toy truck of mass 2 kg. They both stop. What was the speed of the truck?

Work and Energy

Work and Energy

When a **force** makes something move **work** is done. The work done is equal to the energy transferred. Work and energy are measured in joules (J):

work done by a force (J) = force (N) × distance moved by force in direction of the force (m)

When work is done *by* something it loses energy, when work is done *on* something it gains energy.

Kinetic Energy

An object that is moving has **kinetic energy** (**KE**). The energy depends on the mass of the object and on the square of the speed. Doubling the speed gives four times the energy.

Example: An air hockey puck, floating on an air table, is almost frictionless. A force does work on the puck – it pushes it a small distance. Energy is transferred and the kinetic energy of the puck increases – it speeds up. When the force stops the puck moves at a constant speed across the table – its kinetic energy is now constant.

kinetic energy (J) = ½ × mass (kg) × [speed (m/s)]^2

Energy does not have a direction. Speed or velocity can be used to calculate the kinetic energy.

Example: A ball of mass 0.27 kg and speed 3 m/s has KE = 0.5 × 0.27 kg × (3 m/s)^2 = 0.9 J.

Gravitational Potential Energy

Gravitational potential energy (**GPE**) is the stored energy that an object has because of its position above the surface of the Earth.

Example: Doing work – increasing the GPE.

When you lift a 10 N weight (a mass of 1 kg) from the floor to a high shelf, a height difference of 2 m, you have done work on the weight.

Increasing the GPE

The work done = 10 N × 2 m = 20 J and this is equal to the increase in the GPE of the weight.
Change in GPE (J) = weight (N) × vertical height difference (m).

💡 Boost Your Memory

It's the *change* in GPE that depends on the *change* in height.

Build Your Understanding

Change in **GPE = m g h** where *g* = the gravitational field strength (N/kg), *m* = mass (kg) and *h* = height change (m).

KE = ½ m v^2 where *m* = mass (kg) *v* = speed (m/s)

Rollercoasters

When frictional forces are small enough to be ignored the transfer of energy between KE and GPE can be used to calculate heights and speeds.

Example: A car is driven by a trackside motor to the top of a rollercoaster and then freewheels down the slope.

Increase in GPE of car:

= $m g h$ = 1000 kg × 10 N/kg × 45 m = 450 000 J

Assuming there are no friction forces as the train travels down the slope:

Loss of GPE = gain in KE = 450 000 J

Transferring energy from GPE to KE and back again

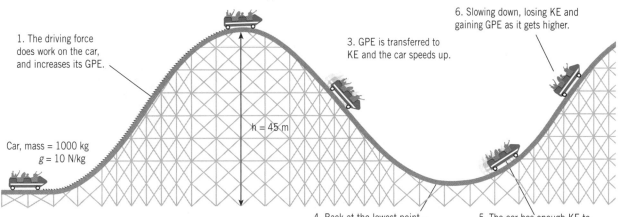

1. The driving force does work on the car, and increases its GPE.
2. The highest point – the car has maximum GPE. It is stopped so KE = 0.
3. GPE is transferred to KE and the car speeds up.
4. Back at the lowest point – maximum KE and GPE = 0.
5. The car has enough KE to continue up the next slope.
6. Slowing down, losing KE and gaining GPE as it gets higher.

Car, mass = 1000 kg
g = 10 N/kg
h = 45 m

Build Your Understanding

Example: To calculate the speed of a car as above:

Gain in KE = 450 000 J

450 000 J = $½ m v^2$ = ½ × 1000 kg × v^2

v^2 = 900 (m/s)² so speed v = 30 m/s

✓ Maximise Your Marks

Remember to square a number by multiplying it by itself. A calculator is useful for finding the square root of a number.

❓ Test Yourself

1. What is the work done by a tractor that pulls a trailer with a force of 1000 N across a 200 m field?
2. A rollercoaster car that has a weight of 12000 N goes to the top of a 30 m slope. What is the gain in GPE?
3. A 900 kg car is travelling at 15 m/s. What is its KE?
4. A toy car rolls down a slope. The gain in KE is less than the loss in PE. Suggest why.

⭐ Stretch Yourself

1. A lift weighs 8000 N it is raised a height of 50 m.
   a) What is the gain in GPE?
   b) What is the work done on the lift by the motor?

Energy and Power

Conservation of Energy

The **Principle of Conservation of Energy** says that the total energy always remains the same. When energy is transferred to the surroundings by heating due to frictional forces it is no longer useful, but it is not lost. We say it has been **dissipated** (spread out) as heat.

The relationship 'gain in KE = loss in GPE' is only true for a falling object if the air resistance (or drag) is small and can be ignored, or if the object is falling in a vacuum.

A skydiver eventually reaches **terminal velocity**. She is still falling, so GPE is being lost, but no KE is being gained. The energy is being used to do work against the frictional force (air resistance). The skydiver and surrounding air will heat up.

Forces when skydiving

Build Your Understanding

The space shuttle, with a lot of KE, needed heat proof tiles to protect it from the heat resulting from doing work against air resistance when it re-entered the Earth's atmosphere.

When a cyclist pedals, but travels at a steady speed, work is done against air resistance and friction. Energy is transferred and heats the bicycle and surroundings. No energy is being transferred as KE to the bicycle unless it speeds up.

Stopping Distances

The distance travelled between the driver noticing a hazard and the vehicle being stationary is called the **stopping distance**:
- Stopping distance = thinking distance + braking distance.
- **Thinking distance** is distance travelled during the drivers reaction time – the time between seeing the hazard and applying the brakes.
- **Braking distance** is the distance travelled while the vehicle is braking.

This diagram shows the shortest stopping distances at different speeds.

When speed doubles:
- Thinking distance doubles.
- Braking distance is four times as far.

The stopping distance increases with speed

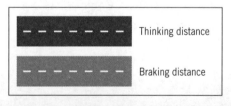

Longer Stopping Distances

The stopping distances are also longer if:
- The driver is tired, affected by some drugs (including alcohol and some medicines), or distracted and not concentrating. Thinking distance is increased.
- The road is wet or icy or the tyres or brakes are in poor condition. The friction forces will be less so the braking distance is increased.
- The vehicle is fully loaded with passengers or goods. The extra mass reduces the deceleration during braking, so the braking distance is increased.

These stopping distances are taken into account when setting road speed limits. Drivers should not drive closer than the thinking distance to the car in front, to allow for time to react. They should reduce speed in bad weather to allow for the increased braking distance.

Braking and Kinetic Energy

- When speed doubles reaction time is the same:
  thinking distance = speed × reaction time
  (The thinking distance doubles).
- Work done by the brakes against friction = loss in KE.

braking force × braking distance = ½ mv^2

$$\text{braking distance} = \frac{\text{mass} \times \text{speed}^2}{2 \times \text{braking force}}$$

(The braking distance is four times as far.)

- The **thinking distance** depends on **speed**.
- The **braking distance** depends on **(speed)^2**.

Example: At three times the speed, braking distance is nine times as far.

Power

Power is the work done, or energy transferred, divided by time. Power is measured in watts (W):

$$\text{power (W)} = \frac{\text{work done or energy transferred (J)}}{\text{time (s)}}$$

Build Your Understanding

Power is the rate of energy transfer.

Example: A 7.5 kW crane lifts a 3000 N weight up a height of 10 m. How long does it take?

$$7.5 \text{ kW} = \frac{3000 \text{ N} \times 10 \text{ m}}{t}$$

$$t = \frac{30000 \text{ J}}{7500 \text{ W}} = 4 \text{ s}$$

💡 Boost Your Memory

Units can help you to remember how things are related and do calculations. A watt is a joule per second, so divide energy (joules) by time (seconds) to get power (watts).

✓ Maximise Your Marks

Don't forget to use the correct units in your calculations. It can help to change everything to the basic units, for example kilowatts to watts. An answer without units is not an answer – it's just a number.

❓ Test Yourself

1. Under what conditions is the 'loss in GPE = gain in KE'?
2. What happens to the GPE when an object falls with terminal velocity?
3. Give an example where stopping distances would be longer than shown in the diagram.

⭐ Stretch Yourself

1. What happens to the energy transferred by a pedalling cyclist when travelling at a steady speed?
2. Calculate the stopping distance of a car when travelling at 90 mph.

Electrostatic Effects

Electric Charge

Electric charge can be **positive** or **negative**. **Electrons** are particles with a negative electric charge. They can move freely through a **conductor**, for example any type of metal, but cannot move through an **insulator**.

Two objects attract each other if one is positively charged and the other is negatively charged. Two objects with similar charge (both positive or both negative) repel.

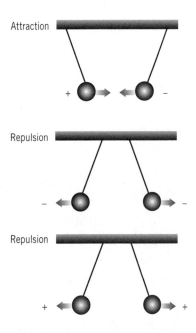

Electrostatic effects are caused by the transfer of electrons. (It is also sometimes called static electricity). When insulators are rubbed electrons are rubbed off one material and transferred to the other.

A polythene rod rubbed with a duster picks up electrons from the duster and becomes negatively charged, leaving the duster positively charged.

A Perspex rod rubbed with a duster loses electrons to the duster and becomes positively charged, leaving the duster negatively charged.

Materials that are positively charged have missing electrons. Materials that are negatively charged have extra electrons.

The insulated rod and the cloth have opposite charge

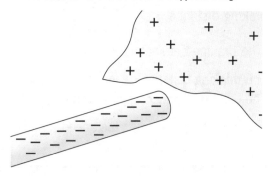

💡 Boost Your Memory

Like charges repel. Unlike charges attract. Remember, opposites attract!

Build Your Understanding

Conductors cannot be charged unless they are completely surrounded by insulating materials, such as dry air and plastic, otherwise the electrons flow to or from the conductor to discharge it.

An *insulated conductor* can be charged by rubbing it with a charged duster, or touching it with a charged rod. Some electrons are transferred, so that the charge is spread out over both objects.

A conductor can be discharged by touching it with another conductor, for example a wire, so that electrons can flow along the wire and cancel out the charge.

✓ Maximise Your Marks

Remember that it is the electrons that move from one object to another.

The Earth Connection

To stop conductors becoming charged they can be **earthed**. A thick metal wire is used to connect them to a large metal plate in the ground. This acts as a large reservoir of electrons. Electrons flow so quickly to, or from, earth that objects connected to earth do not become charged.

Dangerous or Annoying?

The human body conducts electricity. When a large flow of charge affects our nerves and muscles we call this an **electric shock**.

- Small electrostatic shocks are not harmful.
- Larger shocks can be dangerous to people with heart problems because a flow of charge through the body can stop the heart.
- Lightning is a very large electrostatic discharge. When it flows through a body it is often fatal.

Annoying Electrostatic Charge

Charged objects, like plastic cases and TV monitors, attract small particles of dust and dirt.

Clothing can be charged as you move and 'clings' to other items of clothing. Synthetic fibres are affected more than natural fibres as they are better insulators.

If charge builds up on you and you touch metal, for example a car door, the charge flows from you to the metal, and you get a shock.

A **spark** occurs when electrons jump across a gap. This can cause an explosion if there are:

- inflammable vapours like petrol or methanol
- powders in the air, like flour or custard, which contain lots of oxygen – as a dust they can explode
- inflammable gases like hydrogen or methane

Build Your Understanding

Lorries containing flammable gases, liquids and powders are connected to an earth before loading or unloading. Aircraft are earthed before being refuelled. This prevents charge from building up on metal pipes or tanks when the loads are moved, so there is no danger of a spark igniting the load.

Anti-static sprays, liquids and cloths stop the build up of static charge. These work by increasing the amount of conduction, sometimes by attracting moisture because water conducts electricity.

If you stand on an insulating mat, or wear shoes with insulating soles, when you touch a charged object this will reduce the chance of an electric shock because the charge will not flow through you to earth. You become charged and stay charged until you touch a conductor.

❓ Test Yourself

1. Why is a plastic rod attracted to a cloth it has been rubbed with?
2. What particles are transferred when a balloon is rubbed with a cloth?
3. How many types of electric charge are there?
4. Why do you become charged walking on a nylon carpet, but not on a woollen carpet?

⭐ Stretch Yourself

1. When a plastic rod and a Perspex rod are both charged by rubbing they attract.
    a) What does this tell you about the charges on them?
    b) Would you expect two rubbed polythene rods to attract or repel?
2. Why are aircraft earthed before being refuelled?

Uses of Electrostatics

Electrostatic Precipitators

Electrostatic precipitators remove dust or smoke particles from chimneys, so that they are not carried out of the chimney by the hot air:
- Charged metal grids are put in the chimneys.
- The smoke particles pass through the grids and become charged.
- Plates at the side are charged opposite to the grids.
- The smoke particles are attracted and stick to the plates.
- The smoke particles clump together on the plates to form larger particles.
- The plates are struck and the large particles fall back down the chimney into containers.

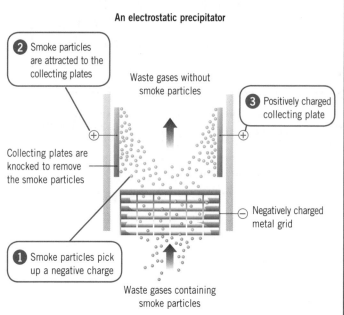

An electrostatic precipitator

1. Smoke particles pick up a negative charge
2. Smoke particles are attracted to the collecting plates
3. Positively charged collecting plate

Collecting plates are knocked to remove the smoke particles

Negatively charged metal grid

Waste gases without smoke particles

Waste gases containing smoke particles

Build Your Understanding

The grids are connected to a high voltage. They attract or repel charges in the smoke particles, so the particles become charged.

The grids are positively charged in some designs and negatively charged in others. If the grids are positively charged, the plates are earthed. The smoke or dust particles lose electrons and become positively charged. They induce a negative charge on the earthed metal plate and are attracted to the plate.

If the grids are negatively charged the plates are positively charged. The smoke or dust particles gain electrons and become negatively charged. They are attracted to the positively charged metal plates.

Paint Spraying

The paint and the object are given a different charge so that the paint is attracted to the object:
- The spray gun is charged so that it charges the paint particles.
- The paint particles repel each other to give a fine spray.
- The object is charged with the opposite charge to the paint.
- The object attracts the paint.
- The paint makes an even coat, it even gets underneath and into parts that are in shadow.
- Less paint is wasted.

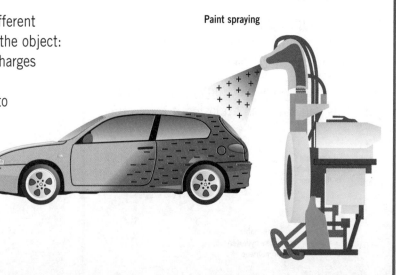

Paint spraying

Defibrillators

When the heart beats the heart muscle contracts. A **defibrillator** is used to start the heart when it has stopped.

The dotted lines show the path of the charge through chest, and heart:
- Two electrodes called paddles are placed on the patient's chest.
- The paddles must make a good electrical contact with the patient's chest.
- Everyone including the operator must 'stand clear' so they don't get an electric shock.
- The paddles are charged.
- The charge is passed from one paddle, through the chest to the other paddle to make the heart muscle contract.

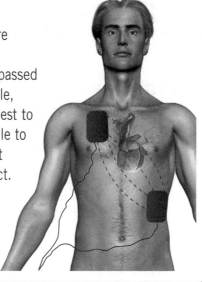

Build Your Understanding

The paddles take a few moments to charge up and then the discharge happens quickly. The electrons move through the heart muscle. Often the heart has not stopped, but has lost its steady rhythm. The defibrillator allows the heart to restart beating to its normal rhythm again.

💡 Boost Your Memory

Remember these uses: Smoke, Sprays, and Shock.

✓ Maximise Your Marks

When you are describing how these applications work, explain what happens to the electrons in each case and how this affects the charge on the objects.

Crop Spraying

Fertiliser and insecticide spray nozzles are charged so that the droplets leaving the nozzle are charged. They repel each other and they are attracted to uncharged objects like the plants. The fine droplets cover the plant better and do not collect into large drops. They are less likely to drift in the wind and get wasted. This means that much less is used which saves money and is better for the environment.

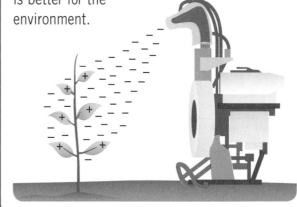

❓ Test Yourself

1. Why do the plates in the electrostatic precipitator have the opposite charge to the grids?
2. What would happen if the paint drops had the same charge as the car body?
3. Give an advantage of electrostatic crop spraying.
4. Why is it important to make sure no one except the patient gets a shock from a defibrillator?

⭐ Stretch Yourself

1. Why are the grids in an electrostatic precipitator connected to a high voltage?
2. Sometimes metal objects are earthed instead of being given a charge. Explain how this works.

Electric Circuits

Circuit symbols

Component	Symbol	Component	Symbol
switch (open)		lamp	
switch (closed)		fuse	
cell		fixed resistor	
battery		variable resistor	
ammeter		light dependent resistor (LDR)	
voltmeter		thermistor	
junction of conductors		diode	
motor		generator	
power supply		a.c. power supply	

💡 Boost Your Memory

To learn the symbols, draw or print them onto cards – one card with the word and one card with the symbol. Then use them to play 'pairs'. Place them face down and turn two over at a time. If you get a 'pair' of the matching word and symbol you keep it. If not you place them face down again. The winner is the person with the most pairs.

✓ Maximise Your Marks

Take care when drawing circuit diagrams. Although the shape of the connecting wires does not matter they must join the components properly – electricity can't flow through gaps. The ammeter and voltmeter symbols are circles, not squares, and the symbol is 'A' not 'a'.

Electric Current

Electric current:
- Is a flow of **electric charge**.
- Only flows if there is a compete circuit. Any break in the circuit switches it off.
- Is measured in **amps** (A) using an **ammeter**.
- Is not used up in a circuit. If there is only one route around a circuit the current will be the same wherever it is measured.
- Transfers energy to the components in the circuit.

A **series circuit** is a circuit with only one route around it. The current measured on each ammeter will be the same.

A **parallel circuit** has more than one path for the current around the circuit. In this circuit there are two paths, marked in red and blue, around the circuit. The current measured on ammeters B and C adds up to give current measured on ammeter A and on ammeter D.

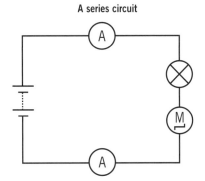

A series circuit

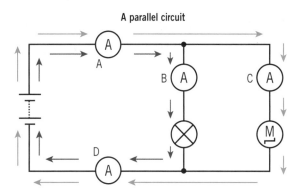

A parallel circuit

Build Your Understanding

Electric current is a flow of positive charge so the direction of the current is opposite to the direction of the electron flow, because electrons are negatively charged.

Electric charge is measured in coulombs (C). The amount of electric charge passing a point in the circuit depends on the current:

Charge (C) = current (A) × time (s)

$Q = I\,t$

Metal conductors contain lots of electrons that are free to move. When the battery makes the electrons move they flow in a continuous loop around the circuit. In insulators there are few charges that are free to move.

Batteries supply direct current, d.c. so the charges always move in the same direction from the positive terminal, around the circuit to the negative terminal. Mains electricity is produced by generators and the charges reverse direction. This is called alternating current (a.c.).

At a junction in a circuit, the total current flowing into the junction must be the same as the total current flowing out of the junction.

? Test Yourself

1. In a parallel circuit, if ammeter B reads 0.3 A and ammeter C reads 0.5 A what is the reading on:
   a) Ammeter A; b) Ammeter D?

2. In a parallel circuit, if ammeter B reads 500 mA and ammeter A reads 900 mA what is the reading on:
   a) Ammeter C; b) Ammeter D?

★ Stretch Yourself

1. If a current of 2 A is switched on for 10 s, how much charge has flowed?

2. In the parallel circuit above, what will always be true about the readings on ammeters A, B and C?

Voltage or Potential Difference

Voltage or Potential Difference

Voltage is also called **potential difference (p.d)**.

Voltage is:
- Measured between two points in a circuit.
- Measured in **volts** (V) using a **voltmeter**.

✓ Maximise Your Marks

Students often confuse voltmeters and ammeters. Always say **voltage across** and **current through**.

This will remind you that to measure the current flowing through a component you must connect the ammeter in line, so that the current flows through it. To measure the voltage across the component you must connect the voltmeter across the component making a connection on either side of it.

The higher the voltage of a battery the higher the 'push' on the charges in the circuit.

This diagram shows how to connect voltmeter A to measure the voltage supplied by the battery, and how to connect voltmeter B to measure the voltage across one of the lamps.

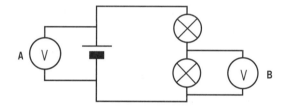

Build Your Understanding

Potential difference, or voltage is a measure of energy transferred to (or from) the charge moving between the two points.

In the diagram above:
- Voltmeter A is measuring the energy transferred *to* the charge.
- Voltmeter B is measuring the energy transferred *from* the charge.

The potential difference (voltage) between two points is the work done (energy transferred) per coulomb of charge that passes between the two points.

Potential difference (V) = $\dfrac{\text{Work done (J)}}{\text{Charge (C)}}$

$$V = \dfrac{W}{Q}$$

💡 Boost Your Memory

Remember that 'a volt is a joule per coulomb'. Add to this that 'a coulomb is an amp second' and you can work out most of the electricity relationships you need.

Voltage in Series or Parallel

When components are connected in **series** the voltage, or p.d., of the power supply is shared between the components.

Adding the measurements on the three voltmeters gives the power supply p.d. When components are connected in **parallel** to a power supply, the voltage, or p.d., across each component is the same as that of the power supply.

The measurements on all the voltmeters are the same.

Voltage in Series and Parallel Circuits

Adding the measurements on the three voltmeters gives the power supply p.d.

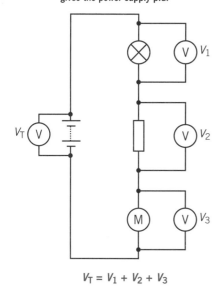

$V_T = V_1 + V_2 + V_3$

The measurements on all the voltmeters are the same

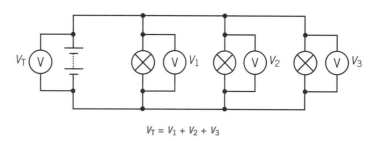

$V_T = V_1 + V_2 + V_3$

Build Your Understanding

Only identical cells should be connected together. In series the p.d. will be the sum of the p.d.s of the cells. In parallel, the p.d. is unchanged. The current will be larger when the cells are in series. In parallel, the current is unchanged, but the cells will last longer because there is more stored charge.

Cells connected in series and in parallel

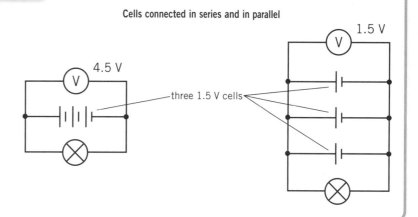

? Test Yourself

1 a) In the series circuit the battery voltage is 12 V, the voltage across the motor is 6 V and across the resistor is 4 V. What is the voltage across the lamp?

b) Does this mean that the current will be different in each component? Explain your answer.

2 a) In the parallel circuit above if the battery voltage is 9 V what is the voltage across each of the lamps?

b) Does this mean that the lamps will be equally bright? Explain your answer.

★ Stretch Yourself

1 If the voltage across a lamp is 9 V and 10 C of charge flows through the lamp how much energy has been transferred?

2 What voltage would be supplied by five 1.5 V cells:

a) In series.

b) In parallel.

3 What advantage is there to connecting the five cells in parallel?

Resistance and Resistors

Resistance

The components and wires in a circuit **resist** the flow of electric charge. When the **voltage**, (or p.d.), *V*, is fixed, the larger the **resistance** of a circuit the less **current**, *I*, passes through it.

The resistance of the connecting wires is so small it can usually be ignored.

Other metals have a larger resistance, for example the filament of a light bulb has a very large resistance. Metals get hot when charge flows through them. The larger the resistance the hotter they get. A light bulb filament gets so hot that it glows.

Resistance is measured in **ohms** (Ω).

Resistance (Ω) = $\dfrac{\text{voltage (V)}}{\text{current (A)}}$

$R = \dfrac{V}{I}$

Build Your Understanding

Metals are made of a lattice of stationary positive ions surrounded by free electrons. The moving electrons form the current. In metals with low resistance the electrons require less of a 'push' (p.d) to get through the lattice. The moving electrons collide with the stationary ions and make them vibrate more. This increase in kinetic energy of the lattice increases the temperature of the metal.

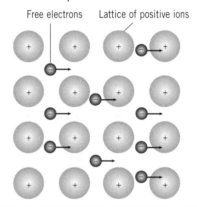

Combining Resistors

Components can be added to a circuit in **series** or in **parallel**.

For components in series:
- Two (or more) components in series have more resistance than one on its own.
- The current is the same through each component.
- The p.d. is largest across the component with the largest resistance.
- The p.d.s across the components add up to give the p.d. of the power supply.
- The resistances of all the components add up to give the total resistance of the circuit.

For components in parallel:
- Two (or more) components in parallel have less resistance than one on its own.
- The current through each component is the same is if it were the only component.
- The total current will be the sum of the currents through all the components.
- The p.d. across all the components will be the same as the power supply p.d.
- The current is largest through the component with the smallest resistance.

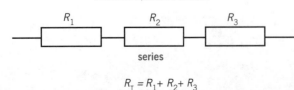

$R_T = R_1 + R_2 + R_3$

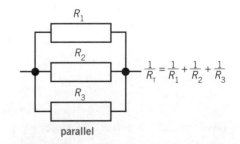

$\dfrac{1}{R_T} = \dfrac{1}{R_1} + \dfrac{1}{R_2} + \dfrac{1}{R_3}$

Build Your Understanding

For components in series:
- Two (or more) components in series have more resistance than one on its own. This is because the battery has to push charges through both of them.
- The p.d. is largest across the component with the largest resistance. This is because more work is done by the charge passing through a large resistance than through a small one.

For components in parallel:
- A combination of two (or more) components in parallel has less resistance than one component on its own. This is because there is more than one path for charges to flow through.
- The current is largest through the component with the smallest resistance. This is because the same battery voltage makes a larger current flow through a small resistance than through a large one.

Boost Your Memory

A series is one after the other, parallel lines are side by side, so series circuits have one component after another and parallel circuits have components that can be drawn side by side.

Fixed Resistors

In some components, such as **resistors** and **metal conductors**, the resistance stays constant when the current and voltage change, providing that the temperature does not change.

For this type of fixed resistance if the voltage is increased the current increases so that a graph of current against voltage is a straight line. The current is **directly proportional** to the voltage – doubling the voltage doubles the current. Components that obey this law (**Ohm's Law**) are sometimes called **ohmic** components.

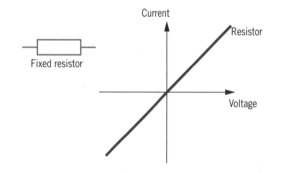

A graph of current against voltage for a resistor

Maximise Your Marks

When there is no voltage there is no current, so graphs of I against V pass through the point (0,0). Remember this when you are drawing graphs.

Test Yourself

1. Lamp voltage = 9 V current = 0.1 A. What is the resistance?
2. Voltage = 12 V resistance = 200 Ω. What is the current?
3. In the series circuit on page 54 R_1 =100 Ω, R_2 = 200 Ω, R_3 =300 Ω.
   a) What is the total resistance?
   b) Which resistor will have the largest voltage across it?

Stretch Yourself

1. Why do metals get hot when an electric current flows through them?
2. A student takes these measurements for a resistor:
   $V = 3$ V; $I = 20$ mA
   $V = 4.5$ V; $I = 30$ mA

   Does the resistor obey Ohm's Law? Explain your answer.

Special Resistors

A Filament Lamp

The wire in a **filament lamp** gets hotter for larger currents. This increases the resistance so the graph of current against voltage is not a straight line.

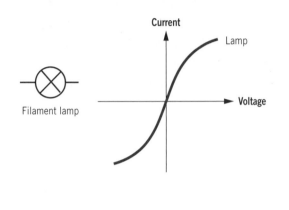

Variable Resistors

A **variable resistor** changes the current in a circuit by changing the resistance. This can be used to change how circuits work, for example to change:
- How long the shutter is open on a digital camera.
- The loudness of the sound from a radio loud speaker.
- The brightness of a bulb.
- The speed of a motor.

Inside one type of variable resistor is a long piece of wire made of metal with a large resistance (called **resistance wire**). To alter the resistance of the circuit a sliding contact is moved along the wire to change the length of wire in the circuit.

Special Resistors

The resistance of a **light dependent resistor (LDR)** decreases as the amount of light falling on it increases. It can be used in a circuit to switch a lamp on, or off, when it gets darker, or lighter.

The resistance of the most common type of **thermistor** (a negative temperature coefficient (NTC) thermistor) decreases as the temperature increases. It can be used in a circuit to switch a heater or cooling fan, on, or off, at a certain temperature.

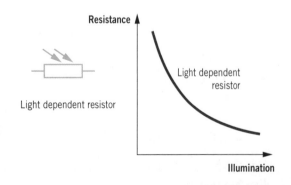

A graph of resistance against intensity of light for a light dependent resistor

Current will only flow through a **diode** in one direction – the forward direction. In one direction its resistance is very low, but in the other direction, called the **reverse direction**, its resistance is very high.

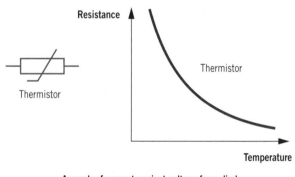

A graph of resistance against temperature for a thermistor

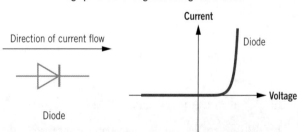

A graph of current against voltage for a diode

Build Your Understanding

Two resistors can be used in a circuit to provide an output p.d. with the value that is wanted from a higher input p.d. This is called a potential divider circuit.

current: $I = \dfrac{V_{in}}{(R_1 + R_2)} = \dfrac{V_1}{R_1} = \dfrac{V_2}{R_2}$

voltage: $V_{in} = V_1 + V_2$

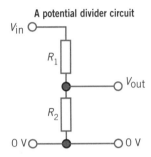

A potential divider circuit

The p.d.s (or voltages) are divided in the same ratio as the resistances.

When a thermistor is used as one of the resistors the resistance will change with the temperature. This circuit will produce a p.d. that changes with temperature and so it can be used to switch a heater on or off.

An LDR can be used in the same way.

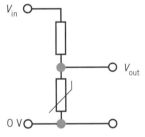

The temperature dependent potential divider

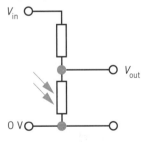

A light dependent potential divider

✓ Maximise Your Marks

When an ordinary resistor gets hotter its resistance increases, but for most common thermistors resistance decreases. When light intensity increases LDRs resistance decreases. The extra energy makes it easier for current to flow in these materials.

Light Emitting Diodes

A **light emitting diode (LED)** is a diode that emits light. LEDs are becoming widely used as low voltage and low energy sources of light.

Notice that symbols for diodes and LEDs (and also LDRs) sometimes include a circle:

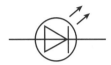

💡 Boost Your Memory

The diode symbols are like arrow heads which show which way the current goes.

❓ Test Yourself

1. Why is the resistance of a filament lamp higher when it is switched on?
2. Why would the metal used to make lamp filaments be unsuitable for connecting components in a circuit?
3. What property of a thermistor makes it useful for controlling a heater?
4. Mains electricity is a.c. and computers require a d.c. supply. Which component would be useful in a circuit to connect a computer to use mains electricity? Why?

⭐ Stretch Yourself

1. In the potential divider circuit if the input p.d. V_{in} = 5 V, R_1 = 300 Ω, R_2 = 200 Ω. What is:
   a) V_1; b) V_2?
2. In the circuit with the thermistor when the temperature increases what happens to:
   a) The resistance of the thermistor?
   b) The p.d. across the thermistor?
   c) The p.d. across the resistor?

The Mains Supply

Safe Use of Mains Electricity

- Mains voltage is 230 V a.c.
- The direction of the current and voltage changes with frequency = 50 Hz
- An electric shock from the mains can kill.

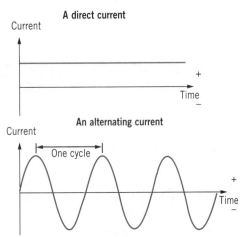

You should never interfere with mains electricity in any way. It should only be a professional electrician. The colour code for mains electricity cables used in buildings and appliances is shown below.

Name of wire	Colour of insulation	Function of the wire
Live	Brown	Carries the high voltage.
Neutral	Blue	The second wire to complete the circuit.
Earth	Green and yellow	A safety wire to stop the appliance becoming live.

A **fuse** is a piece of wire that is thinner than the other wires in the circuit. It will melt first if too much current flows before the wires overheat.

A 3 A fuse will melt if a current of 3 A flows through it. Choose a fuse that is the lowest value, which is more than the normal operating current. If there is a fault, or if too many appliances are plugged into one socket, resulting in a large current, then the fuse will melt and break the circuit preventing a fire.

The **earth wire** is connected to the metal case of appliances so that when they are plugged into the mains supply the metal case is earthed (see page 47). If there is a fault and the live wire touches the metal case, a very large current flows through the low-resistance path to earth melting the fuse wire and breaking the circuit.

Double insulated appliances have cases that do not conduct (usually plastic) and have no metal parts that you can touch, so they do not need an earth wire.

This diagram shows the wiring of a 3-pin plug for a heater with a metal case. The **fuse** is always connected to the brown, **live** wire. A **cable grip** is tightened where the cable enters the plug to stop the wires being pulled out.

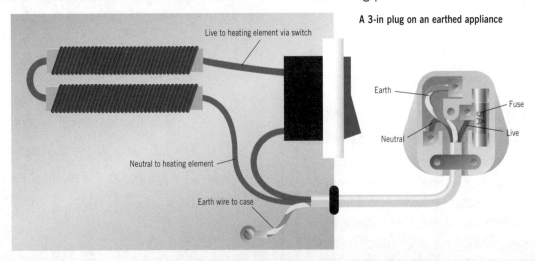

A 3-in plug on an earthed appliance

Build Your Understanding

The **fuse** takes a short time to melt. It will not prevent you from getting an electric shock if you touch a live appliance. A residual current circuit breaker **(RCCB)** is safer.

These are switches to cut off the electricity very quickly if they detect a difference in the current flowing in the live and the neutral wires, (for example, if the current flows through a person, or appliance casing). Another advantage is that they can be switched back on once the fault is fixed, whereas a fuse must be replaced.

An RCCB can be part of mains circuit in a building, or a plug in device that goes between the appliance and the socket.

Appliances that are dangerous include:
- Those where the cable could get wet, or be cut, for example, lawn mowers and power tools.
- Music amplifiers connected to a metal instrument that someone is playing.

Boost Your Memory

Try explaining these safety features to someone. You'll soon find out if you remember them.

Electrical Power

The power is the rate at which the power supply transfers electrical energy to the appliance. It is measured in watts (W).

$$\text{power (W)} = \frac{\text{energy (J)}}{\text{time (s)}}$$

Electrical power (W) = current (A) × voltage (V)

$P = IV$

Electrical energy (J) = current (A) × voltage (V) × time (s)

$E = IVt$

Example: What is the current in a 2.8 kW kettle?

Using $P = IV$

$I = P \div V$

$I = 2800 \text{ W} \div 230 \text{ V} = 12.2 \text{ A}$

Power and Resistance

Another useful equation is:

Using $P = IV$ and $R = \frac{V}{I}$ so $V = IR$

$P = I \times (IR) = I^2R$

Power (W) = [current(A)]² × resistance (Ω)

$P = I^2R$

✓ Maximise Your Marks

Check carefully which equations you are given in the exam and which you need to learn. Make sure you know where to find them on the exam paper. You may find the triangle method useful for rearranging equations.

❓ Test Yourself

1. Sam replaces a fuse with a piece of high resistance wire. Why is this a bad idea?
2. The earth wire is not connected to a metal appliance. Why is this dangerous?
3. What is the current in:
    a) A 2.5 kW kettle.
    b) A 9 W lamp.
    c) A 300 W TV.
4. Fuses come in 3 A, 5 A and 13 A. Which would you use for each appliance in question 3?

★ Stretch Yourself

1. Give two advantages of using an RCCB with outdoor Christmas lights.
2. Cables have 100 Ω resistance. Calculate the power wasted heating the cables when the current is:
    a) 0.5 A
    b) 1 A

Atomic Structure

The Atom

Atoms are about 10^{-10} m or 0.1 nanometres in diameter.

Neutral atoms have the same number of protons and electrons. When ionisation occurs atoms gain or lose electrons, becoming negatively or positively charged **ions**:

- The **atom** is mostly empty space with almost all the mass concentrated in the small positively charged nucleus at the centre.
- The nucleus is very small compared to the volume, or shell, around the nucleus that contains the **electrons**.

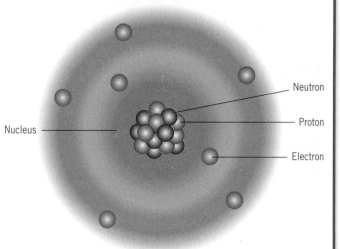

Particles in the atom	Symbol	Where found in the atom	Relative mass	Relative charge
Proton	p	in nucleus	1	+1
Neutron	n	in nucleus	1	0 (neutral)
Electron	e	outside nucleus	$\frac{1}{1840}$	−1

The **nucleus** of an atom contains particles called **protons** and **neutrons**. These are also called **nucleons**.

The **atomic number** or **proton number**, **Z**, is the number of protons in the nucleus. The number of protons is what makes the atom into the element it is, so, for example, hydrogen always has one proton and carbon always has six.

The **mass number** or **nucleon number**, **A**, is the total number of protons and neutrons in the nucleus.

Isotopes of an element have the same number of protons in the nucleus, but different numbers of neutrons.

Isotopes of the same element have exactly the same chemical properties, but they have different mass and nuclear stability.

For example, carbon-12 is a stable isotope of carbon with 6 protons and 6 neutrons and carbon-14 is a radioactive isotope with 6 protons and 8 neutrons.

💡 Boost Your Memory

Make a set of flash cards with these words on one side and what they mean on the other. Keep looking at them and this will help you remember them.

Build Your Understanding

Nuclei are given symbols, for example this is the symbol for the stable isotope carbon-12 which has 6 protons and 6 neutrons: $^{12}_{6}\text{C}$

This is the symbol for an alpha particle (see page 62), which is the same as a helium nucleus. Sometimes α is used instead of He: $^{4}_{2}\text{He}$

A beta particle (see page 62) is not a nucleus, but this symbol is used for a beta particle in a nuclear equation. Sometimes β is used instead of e: $^{0}_{-1}\text{e}$

Nuclear Equations

Before and after a nuclear decay or reaction:
- The total of the mass numbers must be the same.
- The total of the atomic numbers must be the same.

Example: Alpha decay of radon-220.

$^{220}_{86}\text{Rn} \rightarrow {}^{216}_{84}\text{Po} + {}^{4}_{2}\text{He}$

$220 = 216 + 4$ and $86 = 84 + 2$

Example: Beta decay of carbon-14.

$^{14}_{6}\text{C} \rightarrow {}^{14}_{7}\text{N} + {}^{0}_{-1}\text{e}$

$14 = 14 + 0$ and $6 = 7 + (-1)$

A Model of the Atom

Before 1910 scientists had a **plum pudding** model of the **atom.**

The atom is made of positively charged material (pudding) with negatively charged electrons (plums) inside.

Ernest Rutherford investigated the structure of atoms by firing **alpha particles** at **gold foil**.

What happened to the alpha particles	Ernest Rutherford's explanation – the nuclear atom
Most went straight through the foil, without being deflected.	The atom is mostly empty space.
Some were deflected and there was a range deflection angles.	Parts of the atom have positive charge.
A very small number were 'back-scattered' – they came straight back towards the alpha particle source.	There is a tiny region of concentrated mass and positive charge which repels the very small number of alpha particles that have a head-on collision.

Build Your Understanding

The experiment was done in a vacuum, so that the alpha particles were not stopped by the air and it was surrounded by a fluorescent screen, so that a small flash of light was seen when an alpha particle hit the screen.

Hans Geiger and Ernest Marsden counted small flashes of light at different angles for hours.

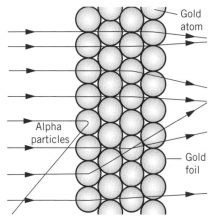

Alpha particles scattering by gold foil

✓ Maximise Your Marks

Remember: Most straight through, some deflected, very few straight back.

❓ Test Yourself

1. Which particles are found in the:
   a) atom; b) nucleus?
2. The atomic number of oxygen is 8. What does this tell you?
3. Which two are isotopes of an element: nitrogen-14, nitrogen-16, oxygen-16?

★ Stretch Yourself

1. Write an equation for:
   a) The alpha decay of uranium-235 (symbol= U, Z = 92) to thorium (symbol = Th).
   b) The beta decay of nitrogen-16 (symbol = N, Z = 7) to oxygen (symbol = O).

Radioactive Decay

Radioactive Emissions

There are three main types of radioactive emissions: **alpha particles**, **beta particles** and **gamma rays**.

Radiation	Ionising	Electric charge	Stopped by...	Affected by electric and magnetic fields?
Alpha (α)	Very strongly	–	A few cm of air. A sheet of paper.	Yes
Beta (β)	Yes	+	A thin sheet of aluminium.	Yes
Gamma (γ)	Only weakly	Neutral	A thick lead sheet. Thick concrete blocks.	No

When alpha and beta particles are emitted the nucleus changes into a different element. When gamma rays are emitted the element does not change.

💡 Boost Your Memory

Write these facts on flash cards to look at and learn. Make your own, because writing them helps you to learn the facts.

Build Your Understanding

Alpha emission is when two protons and two neutrons leave the nucleus as one particle, called an **alpha particle**. It is identical to a helium nucleus.

Beta emission is when a neutron decays to a proton and an electron and the high energy electron leaves the nucleus as a **beta particle**.

Gamma emission is when the nucleus emits a short burst of high-energy **electromagnetic radiation**. The gamma ray has a high frequency and a short wavelength.

Alpha, Beta and Gamma

Alpha particle (α)

Beta particle (β)

Gamma rays (γ)

✓ Maximise Your Marks

Common mistakes are to say:
- 'An alpha particle is a helium *atom*', but it is a helium *nucleus*.
- 'A beta particle is an electron from the *atom*', but it is an electron emitted by a neutron in the nucleus when the neutron turns into a proton.

Radioactive Decay

A radioactive material contains nuclei that are unstable and emit nuclear radiation. This process is called **radioactive decay**. Radioactive decay is **random**. It is not possible to predict when it will happen, or make it happen by a chemical or physical process, for example by heating the material.

A radioactive source contains millions of atoms. The number of radioactive emissions a second depends on two things:
- The type of nucleus – some combinations of protons and neutrons are more stable than others.
- The number of undecayed nuclei in the sample – with double the number of nuclei, on average, there will be double the number of emissions per second.

Over a period of time the **activity** of a source gradually reduces.

The **half-life** of an isotope is the average time taken for half of the nuclei present to decay.

Radioactive Decay (cont.)

Example: Technetium-99m (Tc-99m) decays by gamma emission to Technetium-99 (Tc-99) with a half-life of six hours. After six hours, on average, only half of the Tc-99m nuclei remain. After another six hours, on average, only one quarter are left.

This pattern is the same for all isotopes, but the value of the half-life is different. Carbon-14 has a half-life of 5730 years; some isotopes have a half-life of less than a second.

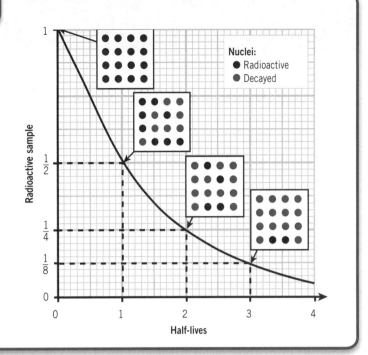

Build Your Understanding

After 10 half-lives, the activity has dropped to less than one thousandth of the original activity. This is often used as a measure of the time for a sample to decay to a negligible amount.

After this number of half-lives...	The activity has dropped to this fraction of the initial value...	Which is the same as...
1	$\frac{1}{2}$	$\frac{1}{2^1}$
2	$\frac{1}{2} \times \frac{1}{2} = \frac{1}{4}$	$\frac{1}{2^2}$
3	$\frac{1}{2} \times \frac{1}{2} \times \frac{1}{2} = \frac{1}{8}$	$\frac{1}{2^3}$
10	$\frac{1}{1024}$	$\frac{1}{2^{10}}$

The number of radioactive emissions a second is called the activity of the source. The activity is measured in bequerel (Bq). An activity of 1 Bq is one radioactive emission per second.

✓ Maximise Your Marks

Common mistakes are to say:
- After three half-lives there are $\frac{1}{3}$ or $\frac{1}{6}$ of the radioactive nuclei left – but it is $\frac{1}{8}$.
- $\frac{1}{8}$ of the nuclei '*have decayed*' – but it is $\frac{7}{8}$ because only $\frac{1}{8}$ are left.

❓ Test Yourself

1. Answer alpha, beta or gamma.
   a) Which has/have a positive charge?
   b) Which is/are stopped by a thin sheet of aluminium?
2. A radioactive source has a half-life of 24 hours. What fraction will remain after a) one day and b) four days?

⭐ Stretch Yourself

1. An alpha particle loses energy and attracts an electron. What has it become?
2. A radioactive source with a half-life of three hours has an activity of 16000 Bq. What is the activity after:
   a) 3 hours; b) 9 hours; c) 30 hours;
   d) When will the activity be 500 Bq?

Living with Radioactivity

Background Radiation

Radioactive materials occur naturally or are man-made. **Cosmic rays** from space make some of the carbon dioxide in the atmosphere radioactive. The carbon dioxide is used by plants and enters food chains. This makes all living things radioactive.

Some rocks are radioactive. We receive a low level of radiation from these sources all the time. It is called **background radiation**. Background radiation comes from:
- Radon (a radioactive gas) from rocks.
- Soil and building materials.
- Medical and industrial uses of radioactive materials.
- Food and drink.
- Cosmic rays (from outer space).
- 'Leaks' from radioactive waste and nuclear power stations.

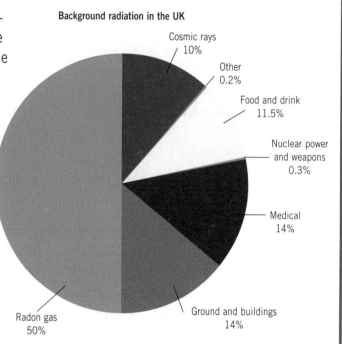

Background radiation in the UK

Build Your Understanding

Some rocks are more radioactive than others, so the level of background radiation depends on the underlying rocks. **Radon** from some rocks builds up in houses. It emits alpha radiation, so it is particularly damaging in the lungs. Houses with high levels of radon can have under-floor fans fitted to keep the radon out of the house.

In some parts of the UK the rocks are more radioactive than in others and there is a higher level of background radiation.

✓ Maximise Your Marks

Background radiation is the name given to radioactive emissions from nuclei in our surroundings. Do not confuse this with radiation from mobile phones or cosmic microwave background radiation.

Sources of background radiation: cosmic rays are from outer space (not the Sun).

Food and drink are radioactive as explained above – not because of food irradiation.

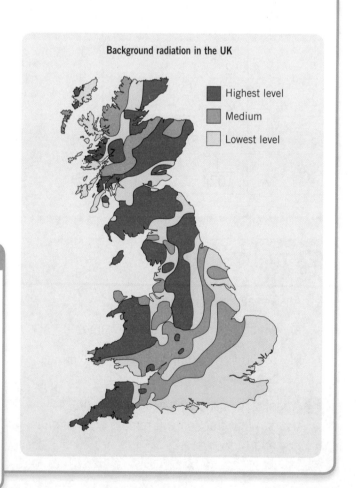

Background radiation in the UK

Dangers of Radiation

Ionising radiation is radiation that has enough energy to charge and ionise the atoms in molecules. The **ions** can take part in chemical reactions. In **living cells** this can damage or kill them. It can damage the **DNA** so that the cell **mutates** into a **cancer** cell.

It is not possible to predict which cells will be damaged by exposure to radiation or who will get cancer.

Scientists studied the survivors of incidents where people were exposed to ionising radiation. They measured the amount of exposure and recorded how many people later suffered from cancer. The risk of cancer increases with increased exposure to radiation. People tend to overestimate the risk from radiation because it is invisible and unfamiliar. They underestimate the risk of familiar activities, like smoking.

Contamination and Irradiation

There are two types of danger from radioactive materials:
- **Irradiation** is exposure to radiation from a source outside the body.
- **Contamination** is swallowing, breathing in, or getting radioactive material on your skin.

A short period of irradiation is not as dangerous as being contaminated because, once contaminated, a person is continually being irradiated.

Alpha radiation has very short range. Even if it reaches the skin it does not penetrate, so there is little danger from irradiation. But it is strongly ionising, so contamination by an alpha source, for example breathing in radon gas, can be very dangerous. Once inside the lungs it will keep irradiating sensitive cells.

Beta radiation has longer range and penetrates the skin so there is more danger from irradiation. It is not as strongly ionising as alpha, so contamination is not as dangerous as with alpha radiation.

Gamma radiation is long range and passes right through the body. There is a danger from irradiation if the radiation levels are high. It is only weakly ionising and most of the rays pass right through the body without hitting anything so contamination is less dangerous than with alpha or beta radiation.

Build Your Understanding

Radiation dose, measured in **sieverts** (Sv), is a measure of the possible harm done to the body. Radiation dose depends on the type of radiation, the time of exposure, and how sensitive the tissue exposed is to radiation. The dose is linked to the risk of cancer developing. Alpha radiation is strongly ionising and so the dose is 20 times larger from alpha than from beta or gamma radiation.

The normal UK background dose is a few milliSieverts. A fatal dose is between 4 Sv and 5 Sv given in one go.

Test Yourself

1. How much of the background radiation in the UK is from radon gas?
2. Give one effect of ionising radiation on living cells.
3. 'She has been irradiated, she will get cancer'. What is wrong with this statement?

Stretch Yourself

1. Name a part of the UK that has high background radiation.

Uses of Radioactive Materials

Choosing the Best Isotope

For each use of radioactive materials:
- Alpha, beta or gamma radiation is chosen depending on the **range** and the **absorption** required.

An isotope is chosen depending on:
- whether it emits alpha, beta or gamma radiation
- how long it remains radioactive, which depends on the half-life.

Uses of Radioactive Isotopes

Smoke detectors
Isotopes which emit **alpha radiation** are used in **smoke detectors**. The alpha radiation crosses a small gap and is picked up by a detector. If smoke is present, the alpha radiation is stopped by the smoke particles. No radiation reaches the detector and the alarm sounds. Beta and gamma radiation are unsuitable because they pass through the smoke.

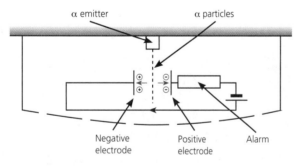

Tracers
Isotopes which emit beta radiation or gamma radiation are used as tracers. Because a tracer is radioactive, detectors can track where it goes. A tracer is added to sewage as it enters a river, to trace its movement. The isotope used has an activity that will fall to zero quickly after the test is done.

Medical and Health Uses

Medical Tracers
Isotopes which emit **gamma radiation** (or sometimes **beta radiation**) are used in **medical tracers**. The patient drinks, inhales, or is injected with the tracer which is chosen to collect in the organ doctors want to examine.

Example: Radioactive iodine is taken up by the thyroid gland, which can then be viewed using a **gamma camera** that detects the gamma radiation passing out of the body. The tracer must not decay before it has moved to the organ being investigated, but it must not last so long that the patient stays radioactive for weeks afterwards.

Treating cancer
Isotopes which emit a higher dose of **gamma radiation** than tracers are used to build up in the cancer and kill the cancer cells.

Alternatively, beams of **gamma rays** are concentrated on a tumour to kill the cancer cells.

Example: Cobalt-60 emits high energy gamma rays and remains radioactive for years.

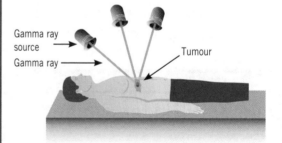

Sterilisation
Gamma radiation destroys microbes and is used for:
- Sterilising equipment, for example surgical instruments.
- Food irradiation to extend the shelf-life of perishable food.

The food or equipment does not become radioactive because it is only **irradiated**, there is no **contamination**. It is also irradiated inside the plastic packaging, so that it stays sterile.

Build Your Understanding

When radioactive materials are used, we have to decide whether the benefits outweigh the risks.

Example: Treatment benefits patients, but does not benefit hospital staff who work with radioactive materials regularly.

Radiation workers have their exposure monitored and take safety precautions to keep their dose as low as possible:
- They wear protective clothing.
- They keep a long distance away, for example they use tongs to handle sources.
- They keep their exposure time short.
- Sources should be shielded and labelled with the radioactive symbol.

Boost Your Memory

Use a mnemonic that contains the first letters of each to remember safety precautions:

My **c**urtains **d**im **t**he **b**right **l**ight

Monitor, **c**lothing, **d**istance, **t**ime, **b**arrier, **l**abel.

Dating

The amount of radioactive carbon left in old materials that were once living can be used to calculate their age.

Some rocks contain a radioactive isotope of uranium that decays to lead, so they can be dated by comparing the amounts of uranium and lead. The more lead there is the older the rock is.

Build Your Understanding

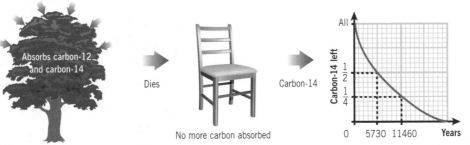

Carbon dating:
- Can only be used for things that once lived.
- Cannot be used for objects older than 10 half-lives = 10 × 5730 years or less than 100 years.

Maximise Your Marks

Learn the properties of alpha, beta and gamma radiation, so that you can explain why each is chosen for different areas.

Test Yourself

1. Give a use of:
   a) Alpha radiation.
   b) Beta radiation.
   c) Gamma radiation.
2. Explain why healthy tissue is not killed by the gamma rays from the cobalt-60 during cancer treatment.

Stretch Yourself

1. Why is alpha radiation used in smoke detectors?
2. Why is carbon dating not used to confirm the age of a 60 000 year old egg?

Nuclear Fission and Fusion

Nuclear Fission

Nuclear fission is when a nucleus splits into two nuclei of about equal size, and two or three neutrons.

Example: After uranium-235 absorbs a neutron:

A neutron absorbed by a uranium nucleus causes a nuclear fission

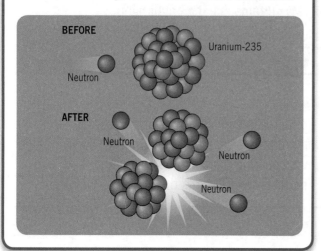

Build Your Understanding

A small amount of mass is converted into a large amount of energy. The energy is calculated using Einstein's equation:

Energy (J) = mass (kg) [speed of light in a vacuum (m/s)]^2

$E = mc^2$

About a million times more energy is released than in a chemical reaction.

Nuclear Power

In nuclear reactors:
- The **fuel rods** are made of **uranium-235** or **plutonium-239**.
- The **moderator** is a material that slows down the neutrons, so they can be absorbed.
- The energy heats up the reactor core.
- A **coolant** is circulated to remove the heat.
- The hot coolant is used to heat water to steam, to turn the power station turbines.
- The **control rods** are moved into the reactor to absorb neutrons to slow or stop the reaction.

Reactors produce **radioactive waste**. The half-lives of some isotopes are thousands of years, so radioactive waste must be kept safely contained for thousands of years.

A nuclear reactor

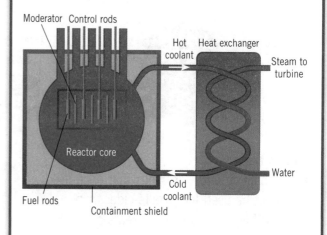

A Chain Reaction

After the new neutrons slow down, they can strike more uranium nuclei and cause more fission events. This is called a **chain reaction**.

If the chain reaction runs out of control it is an **atomic bomb**. In a **nuclear reactor** the process is controlled. In a nuclear power station the energy released is used to generate electricity.

A chain reaction

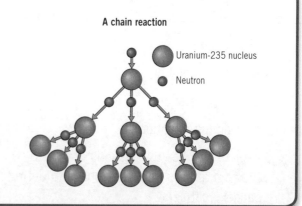

Radioactive Waste

Types of radioactive waste	Examples	Disposal
Low level waste	Used protective clothing	Sealed into containers. Put into landfill sites.
Intermediate level waste	Material from reactors	Mixed with concrete. Stored in stainless-steel containers.
High level waste	Used fuel rods	Kept under water in cooling tanks (it decays so fast it gets hot). Eventually becomes intermediate level waste.

Build Your Understanding

Where is the safest place to store the radioactive waste?
- At the bottom of the sea, but the containers may leak.
- Underground, but the containers may leak and earthquakes or other changes to the rocks may occur.
- On the surface, but needs guarding from terrorists, for thousands of years.
- Blast into space, but there is a danger of rocket explosion.

Nuclear Fusion

When two small nuclei are close enough together they can **fuse** together to form a larger nucleus. This releases a large amount of energy. The problem is getting the two nuclei close enough, because nuclei are positively charged and **repel** each other. Inside stars the temperatures are high enough for the nuclei to have enough energy for nuclear fusion to occur.

Scientists are trying to control the reaction and design nuclear fusion reactors.

More About Fusion

The protons and neutrons inside the nucleus are held together by a force called the **strong force**.

The problem is getting, and keeping, the temperature and pressure high enough to overcome the repulsive force, so that the nuclei get close enough for the strong force to take over. When fusion happens a small amount of mass is converted into a large amount of energy ($E=mc^2$).

💡 Boost Your Memory

Draw up a table to compare fusion and fission. It will help you when you revise.

✓ Maximise Your Marks

You may be asked for advantages and disadvantages, for example, of methods of waste disposal or nuclear reactors. To gain full marks you must give at least one advantage *and* one disadvantage.

If you are asked for your opinion it doesn't matter if you say 'yes' or 'no', but you must say one or the other and give a reason. Remember that most of the marks are for justifying your choice.

❓ Test Yourself

1. What is the difference between nuclear fission and nuclear fusion?
2. Describe a chain reaction in plutonium-239.
3. Explain what control rods are used for.
4. How are fuel rods disposed of when the fuel is used up?

⭐ Stretch Yourself

1. How much energy results from 0.1 g of fuel being converted to energy?
2. What is the force that holds protons and neutrons together called?

Stars

The 'Lifetime' of a Star

Formation:
- Stars begin as large clouds of dust, hydrogen and helium called an **interstellar gas cloud** or a **nebula**.
- Gravity makes the nebula contract, this makes it heat up and it is now a **protostar**.
- When it is hot enough **hydrogen nuclei fuse** together to form **helium nuclei**. This is called **nuclear fusion**. Energy is released as light and other electromagnetic radiation. A **star** has formed.

Stable lifetime, fusing hydrogen:
- The star is one of a large number of **main sequence stars** like our Sun. Stars spend a long time fusing hydrogen. Our Sun will do this for ten thousand million years.
- What happens when a star has fused most of its hydrogen depends on the mass of the star.

Final stages of a small star like our Sun:
- The star cools, becoming redder and expands to form a **red giant**. The core contracts and helium nuclei fuse to form carbon, oxygen and nitrogen.
- After all the helium has fused the star contracts and the outer layers are lost. As these outer layers move away they look to us like a disc that we call a **planetary nebula**.
- The remaining core becomes a small, dense, very hot **white dwarf**. The star will then cool over a very long time and become a **brown or black dwarf**.

Final stages of a massive star:
- The star cools and expands to form a **red supergiant**. The core contracts and **fusion** in the core forms carbon and then heavier elements up to the mass of iron.
- When the nuclear fusion reactions are finished the star cools and contracts, which heats it again until it explodes. The explosion is called a **supernova** and is the largest explosion in the Universe. All the elements heavier than iron that exist naturally on planets were created in supernova explosions.
- The core is left as a **neutron star**. It is very dense. It has a large mass, but a very small volume. If the star is very massive, the core is left as a **black hole**.

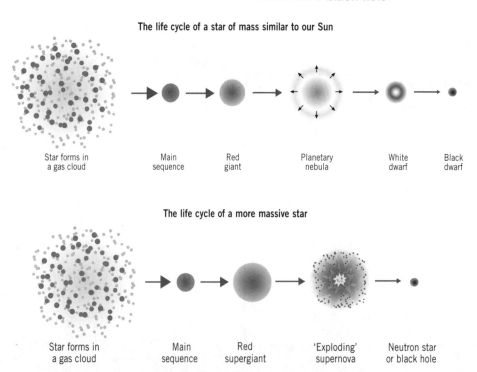

The life cycle of a star of mass similar to our Sun

Star forms in a gas cloud → Main sequence → Red giant → Planetary nebula → White dwarf → Black dwarf

The life cycle of a more massive star

Star forms in a gas cloud → Main sequence → Red supergiant → 'Exploding' supernova → Neutron star or black hole

Build Your Understanding

The particles of dust and gases in the interstellar gas clouds are attracted by gravitational forces between the particles. This is why the cloud contracts. As the core gets hotter the atoms collide at high speed, losing their electrons. The temperature has to be high enough to force the nuclei close together for **fusion** to occur. When light nuclei fuse they release energy.

In a main sequence star the high pressure in the core is balanced by the gravitational forces. The largest stars fuse hydrogen the most quickly so, surprisingly, more massive stars have shorter lifetimes.

Only the most massive stars can cause larger nuclei to fuse, forming elements like magnesium and silicon, but even they cannot form nuclei more massive than iron.

Black holes are so dense that even light cannot escape from their strong gravitational fields.

✓ Maximise Your Marks

Learn the names of all the different stages in the 'lifetime' of a star and make sure that you can use them correctly to describe what happens at each stage.

💡 Boost Your Memory

To remember the stages in the lifetime of a star use a mnemonic, for example:

New **p**lay **m**ates **s**ay **r**ed **g**iants **p**lay **n**icely **w**ith **d**warfs.

Nebula, **p**rotostar, **m**ain sequence, **r**ed **g**iant, **p**lanetary **n**ebula, **w**hite **d**warf.

How we know about Stars

All the information we have about objects outside the Solar System comes from observations made with telescopes. What we know depends on the **electromagnetic radiation** from the stars and galaxies.

All around the Universe there are new stars being formed, **main sequence stars** in the stable part of their 'lifetime' and older stars coming to the end of their time as a star. By observing all of these scientists have worked out the 'life history' of stars.

To find out what stars are made of scientists look at the **spectral lines** in the spectrum of radiation from a star. These are used to identify the elements present in the star.

❓ Test Yourself

1. What is a planetary nebula?
2. When our Sun has completed the whole 'lifetime', what will be left at the end?
3. Which is brighter, a new main sequence star, a new white dwarf or a new supernova?
4. What is the difference between a red giant and a red supergiant star?

⭐ Stretch Yourself

1. How were elements like the carbon and oxygen in your body formed?
2. Why is a black hole black?

Practice Questions

Complete these exam-style questions to test your understanding. Check your answers on pages 125-126. You may wish to answer these questions on a separate piece of paper.

1 The graph below shows speed against time for a car journey.

a) Describe the motion of the car during the journey. (3)

b) What is:

   i) The acceleration during the first 10 seconds?

   ii) The speed after 20 s?

   iii) The acceleration in the last 10 seconds? (2)

c) What does the shaded area represent? (1)

d) How far does the car travel:

   i) In the first 10 seconds?

   ii) In the first 30 seconds? (2)

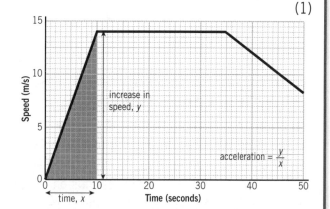

2 Refer to the diagram opposite. Calculate:

a) The current through resistor C. (1)

b) The voltage across resistor C. (1)

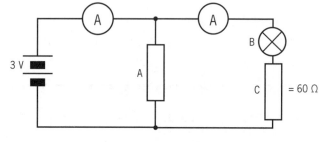

c) The voltage across lamp B. (1)

d) The current through resistor A. (1)

e) The resistance of A. (1)

3 An electric shower has a power of 8.0 kW when it is connected to the 230 V mains supply.

a) Calculate the current in the heater. (1)

b) Explain why very thick cables are needed to connect it to the mains supply. (1)

4 Answer the questions below.

  a) Sodium-24 is a radioactive isotope of sodium that emits gamma radiation. The stable form of sodium is sodium-23. Explain how the nuclear structure of sodium-24 is different to sodium-23. (2)

  ....

  The activity of a sodium-24 source was measured over two days and this graph was plotted.

  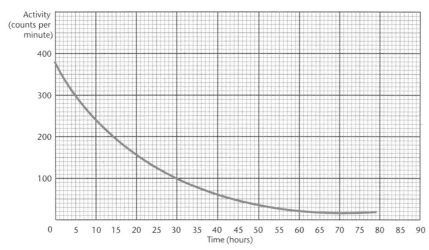

  b) Use the graph to work out the half-life of sodium-24. (1)

  ....

  c) Use the graph to estimate the background activity at the site of the experiment. (1)

  ....

  d) Explain whether a salt of sodium-24 would be a suitable isotope to use for:

  i) The source in a smoke detector. (2)

  ....

  ii) A tracer to find the leak in an underground water pipe.

  ....

5 Anna has a tumour of the thyroid gland. The thyroid gland absorbs iodine. Anna can be cured with an injection of radioactive iodine-131. This is a beta emitter with a half-life of 8 days. Write a patient leaflet for patients like Anna explaining how the treatment works and what she should take into account when deciding whether to have the treatment. (In your answer, aim to write between about 8 and 12 lines.) (6)

....

....

How well did you do?

| 0–8 Try again | 9–16 Getting there | 17–22 Good work | 23–28 Excellent! |

Atomic Structure

Structure of the Atom

An **atom** has a very small, central **nucleus** that is surrounded by shells of **electrons**. The nucleus is found at the centre of the atom. It contains **protons** and **neutrons**:

- Protons have a mass of 1 **atomic mass unit** (**amu**) and a charge of 1+.
- Neutrons also have a mass of 1 amu, but no charge.
- Electrons have a negligible mass and a charge of 1–.

All atoms are **neutral**, therefore there is no overall charge so the number of protons must be equal to the number of electrons.

The **mass number** is the number of protons added to the number of neutrons. The **atomic number** is the number of protons (so it is also known as the **proton number**). All the atoms of a particular element have the same number of protons, for example carbon atoms always have six protons. Atoms of different elements have different atomic numbers.

Sodium has an atomic number of 11, so every sodium atom has 11 protons.

A sodium atom has no overall charge, so the number of electrons must be the same as the number of protons. Sodium atoms, therefore, have 11 electrons.

The number of neutrons is given by the mass number minus the atomic number. In sodium that is 23 – 11 = 12 neutrons.

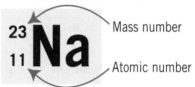

Mass number and atomic number

Electrons occupy the lowest available **shell** (or energy level). This is the shell that is closest to the nucleus. When this is full the electrons start to fill the second shell and so on. The first shell may contain up to two electrons, while the second shell may contain up to eight electrons.

The electron structure of an atom is important because it determines how the atom (and, therefore, the element) will react.

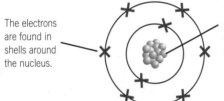

Structure of an atom

The electrons are found in shells around the nucleus.

The nucleus is found at the centre of the atom and contains neutrons and protons.

✓ Maximise Your Marks

To get an A*, you need to be able to recall the radius and mass of a typical atom.

Atoms are very small. They have a radius of about 10^{-10} m and a mass of about 10^{-23} g.

Elements

A substance that is made of atoms with the same atomic number is called an **element**. Elements cannot be broken down chemically. Atoms of different elements have different properties. About 100 different elements have been discovered. The elements can be represented by **symbols**. Approximately 80 per cent of the elements are **metals**. Metals are found on the left-hand side and in the centre of the periodic table. The **non-metal** elements are found on the right-hand side of the periodic table. Elements with **intermediate properties**, such as germanium, are found in group 4.

Build Your Understanding

Elements in the same group of the periodic table have similar chemical properties because they have the same number of electrons in their outer shells.

Magnesium
- Number of protons = 12
- Number of electrons = 12
- Electronic structure = 2, 8, 2

Magnesium is in group 2 of the periodic table because it has two electrons in its outer shell. It is in period 3 of the periodic table because it has three shells of electrons.

Across a period each element has one extra proton in its nucleus and one extra electron in its outer shell of electrons. This means an electron shell is filled with electrons across a period.

Magnesium is in group 2

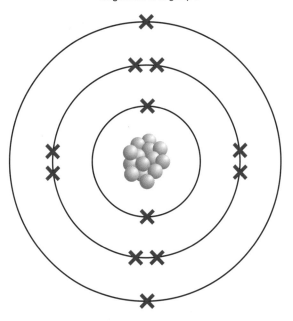

Isotopes

Isotopes of an element have the same number of protons, but a different number of neutrons. So, they have the same **atomic number** but a different **mass number**.

The Isotopes of Chlorine

Chlorine has two isotopes:

- 17 protons
- 17 electrons
- 18 neutrons

$^{35}_{17}Cl$

- 17 protons
- 17 electrons
- 20 neutrons

$^{37}_{17}Cl$

These isotopes will have slightly different physical properties, but will react identically in chemical reactions because they have identical numbers of electrons.

Learn the definition for **relative atomic mass**.

The relative atomic mass of an element compares the mass of atoms of the element with the carbon-12 isotope. The existence of isotopes means that some elements have relative atomic masses that are not a whole number, for example chlorine has a relative atomic mass of 35.5:

- 25 per cent of chlorine atoms have an atomic mass of 37.
- 75 per cent of chlorine atoms have an atomic mass of 35.
- This gives an average atomic mass of 35.5.

❓ Test Yourself

1. What does the nucleus of an atom contain?
2. Which particles are found in shells around the nucleus?
3. What is the charge and mass of a proton?
4. What is the charge and mass of an electron?

⭐ Stretch Yourself

1. Calcium and magnesium both belong to group 2 of the periodic table. Why does the element calcium react in a similar way to the element magnesium?

Atoms and the Periodic Table

The History of the Atom

Ideas about atoms have changed over time as more evidence has become available. Scientists look at the evidence that is available and use this to build a model of what they think is happening.

As new evidence emerges they re-evaluate the model. If the model fits with the new evidence they keep it. If the model no longer works they change it.

Build Your Understanding

John Dalton

In the early 1800s, John Dalton developed a theory about atoms, which included these predictions:

- Elements are made up of small particles called atoms.
- Atoms cannot be divided into simpler substances.
- All atoms of the same element are the same.
- Atoms of each element are different from atoms of other elements.

J.J. Thomson

Between 1897 and 1906 Thomson discovered that atoms could be split into smaller particles. He discovered electrons and found that they:

- Have a negative charge.
- Are very small.
- Are deflected by magnetic and electric fields.

He thought that atoms consisted of tiny negative electrons surrounded by a 'sea' of positive charge. Overall, the atom was neutral. This was called the plum-pudding theory of atoms.

J.J. Thomson's 'plum-pudding' model of the atom

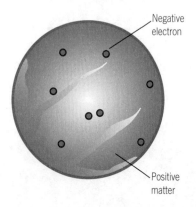

Ernest Rutherford

In 1909, Rutherford examined the results of Geiger and Marsden's experiment in which they had bombarded a very thin sheet of gold with positive alpha particles. The scientists recorded the pathway of the alpha particles through the gold. Rutherford found that while most alpha particles passed through the gold atoms undeflected, a small number of alpha particles were deflected a little and a tiny number of particles were deflected back towards the source. He concluded that the positive charge in the atom must be concentrated in a very small area of the atom. This area is the nucleus of the atom.

The Geiger–Marsden experiment helped Rutherford devise his 'nuclear' model of the atom

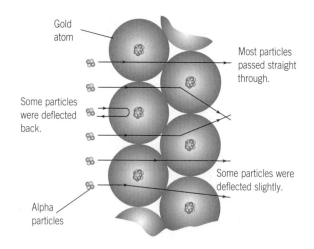

Neils Bohr

In 1913, Neils Bohr deduced that electrons must be found in certain areas in the atom otherwise they would spiral in towards the nucleus.

Atoms and the Periodic Table

Today, scientists consider the periodic table an important summary of the structure of atoms. The periodic table can be used to source the boiling point or density of elements. A detailed periodic table can be used to find the names, symbols, relative atomic masses and atomic number of any element.

In the periodic table, elements are arranged in order of increasing atomic number. It is called a periodic table because elements with similar properties occur at regular intervals or 'periodically'. The elements are placed in horizontal rows, called **periods**, and elements with similar properties appear in the same vertical column. These vertical columns are called **groups**. The elements in group 1 of the periodic table include lithium, sodium and potassium. All the elements in group 1 of the periodic table share similar properties: they are all metals except H (hydrogen) and they all consist of atoms that have just one electron in their outer shell. When these metals react they form **ions**, which have a 1+ charge. Elements in the same period have the same number of shells of electrons.

All the isotopes of an element have the same number of electrons and protons. All the isotopes of an element appear in the same place on the periodic table.

✓ Maximise Your Marks

The modern periodic table is arranged in order of increasing atomic number. You must make this clear in your answers. Many students refer to increasing mass number, which is not correct.

The periodic table

Group	1	2										3	4	5	6	7	0
Period																	
1							H										He
2	Li	Be										B	C	N	O	F	Ne
3	Na	Mg										Al	Si	P	S	Cl	Ar
4	K	Ca	Sc	Ti	V	Cr	Mn	Fe	Co	Ni	Cu	Zn	Ga	Ge	As	Se	Br	Kr
5	Rb	Sr	Y	Zr	Nb	Mo	Tc	Ru	Rh	Pd	Ag	Cd	In	Sn	Sb	Te	I	Xe
6	Cs	Ba	La	Hf	Ta	W	Re	Os	Ir	Pt	Au	Hg	Tl	Pb	Bi	Po	At	Rn
7	Fr	Ra	Ac															

Key: = non-metals, = metals

? Test Yourself

1. Why do scientists have to re-evaluate existing models?
2. How are the elements arranged in the modern periodic table?
3. What are **a)** the horizontal rows and **b)** the vertical columns in the periodic table called?

★ Stretch Yourself

1. John Dalton made a number of predictions about atoms. Today, which of his predictions is not thought to be correct? Explain your answer.

The Periodic Table

Early Ideas

As new elements were discovered, scientists struggled to find links between them.

In 1829, the German chemist Johann Wolfgang Dobereiner noticed that many elements could be put into groups of three, which he called **triads**. If these elements were placed in order of **atomic weight**, the middle element was about the average of the lighter and the heavier element.

He noticed a similar pattern when he compared the densities of the members of a triad. Unfortunately, this pattern only appeared to work some of the time.

History of the Periodic Table

In 1863, the English chemist **John Newlands** noticed that if the known elements were placed in order of their atomic weight, and then put into rows of seven, there were strong similarities between elements in the same vertical column.

This pattern became known as **Newlands' law of octaves**. It was useful for some of the elements, but unfortunately Newlands' pattern broke down when he tried to include the **transition elements**.

In 1869, the Russian chemist **Dimitri Mendeleev** produced his periodic table of elements. In his table, elements with similar properties occurred periodically and were placed in vertical columns called groups.

Like Newlands, Mendeleev arranged the elements in order of increasing atomic weight, but unlike Newlands he did not stick strictly to this order. He left gaps for elements that had yet to be discovered, such as germanium and gallium, and made detailed predictions about the physical and chemical **properties** these elements would have.

Eventually, when these elements were discovered and their properties analysed, scientists confirmed Mendeleev's predictions. His table went from being an interesting curiosity to a useful tool for understanding how a particular element would behave.

By leaving gaps and swapping the order of the elements, Mendeleev had actually arranged the elements in order of increasing atomic number (or the number of protons in the nucleus of an atom), even though protons themselves were not discovered until much later.

In fact, electrons, protons and neutrons were all discovered in the early 20th century.

Build Your Understanding

The Noble Gases

All the noble gases have similar properties (they all have single atoms), so they are in the same group in the periodic table. The noble gas group of elements include helium (He), neon (Ne), argon (Ar), krypton (Kr), xenon (Xe) and radon (Rn). The noble gases were only discovered in the 1890s when chemists noticed that the density of nitrogen made in reactions was slightly different to the density of nitrogen obtained directly from the air. The chemists thought the air might contain small amounts of other gases and so they devised experiments that eventually confirmed the presence of the very unreactive noble gases. They found that the air includes nitrogen, oxygen, neon and argon.

Transition Metals

Transition metals are found in the central block of the periodic table. The transition metals are much less reactive than group 1 metals.

Transition metals have high melting points so, with the exception of mercury, are solid at room temperature. They are hard and strong and make useful structural materials. They do not react with water or oxygen as vigorously as group 1 metals, although many will show signs of corrosion over long periods of time.

Transition metals are used in making a range of useful products such as aeroplanes and cars.

Uses of transition metals

Build Your Understanding

When transition metals form compounds the transition metal ions have variable charges. For example, in copper (II) oxide, CuO, the copper ions have a 2+ charge, while in copper (I) oxide, Cu_2O, the copper ions have a 1+ charge.

The roman numerals given in the name of the transition metal compound shows the charge on the transition metal ion. Transition metal compounds are coloured; group 1 and 2 metal compounds are white:

- Sodium is a group 1 metal.
- Sodium chloride is a white solid.

The compound sodium chloride

Transition metals and transition metal compounds are useful catalysts (chemicals that speed up chemical reactions). Iron is used in the Haber process (which produces ammonia) while nickel is used in the hydrogenation of unsaturated hydrocarbons.

Copper is a transition metal. Hydrated copper (II) sulfate is a blue solid. 'Hydrated' means that the compound contains water.

The compound hydrated copper (II) sulfate

❓ Test Yourself

1. What is the name given to Mendeleev's way of arranging the elements?
2. Why did Mendeleev not include the element germanium in his arrangement of the elements?

⭐ Stretch Yourself

1. What is the charge on the copper ion in the compound copper (II) oxide?

Chemical Reactions and Atoms

Symbols

Each element has its own unique symbol that is recognised all over the world.

Each symbol consists of one or two letters and is much easier to read and write than the full name.

In some cases the symbol for an element is simply the first letter of the element's name. This letter must be a capital letter: the element iodine is represented by the symbol I.

Occasionally, an element may take its symbol from its former Latin name. When this happens, the first letter is a capital and the second letter, if there is one, is lower case: the element mercury is represented by the symbol Hg. This comes from the Latin name for mercury, which was *hydrargyrum*, or **liquid silver**.

Several elements have names that start with the same letter. When this happens, the first letter of the element's name is used, together with another letter from the name. The first letter is a capital and the second letter is lower case: the element magnesium is represented by the symbol Mg.

✓ Maximise Your Marks

Remember to use the periodic table to check any symbols you are using. Don't forget that if a symbol has two letters, the first letter is a capital and the second is lower case.

Chemical Formulae

Compounds consist of two or more different types of atom that have been chemically combined.

A compound can be represented using a chemical **formula** that shows the type and ratio of the atoms that are joined together in the compound.

Ammonia has the chemical formula NH_3. This shows that in ammonia, nitrogen and hydrogen atoms are joined together in the ratio of one nitrogen atom to three hydrogen atoms.

You should take care when writing out the symbols for chemical compounds as some of them are very similar to elements. For example:

- The element carbon has the symbol C.
- The element oxygen has the symbol O.
- The element cobalt has the symbol Co.

The formula CO shows that a carbon atom and an oxygen atom have been chemically combined in a 1 : 1 ratio. This is the formula of the compound carbon monoxide.

The symbol Co represents the element cobalt. Notice how the second letter of the symbol is written in lower case. If it wasn't, it would be a completely different substance.

The formula CO_2 shows that carbon and oxygen atoms have been chemically combined in a 1 : 2 ratio. This is the formula of the compound carbon dioxide.

You need to be very careful when you write chemical symbols and formulae.

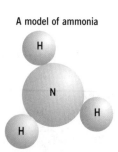

A model of ammonia

✓ Maximise Your Marks

Take care when writing formulae with subscript numbers. They will need to be perfect to get the mark awarded in the exam.

Chemical Reactions

Atoms can join together by:
- **Covalent bonding** – sharing pairs of electrons.
- **Ionic bonding** – giving and taking electrons.

Compounds formed from metals and non-metals consist of ions. These compounds are held together by strong ionic bonds. Compounds formed from non-metals often consist of molecules. The atoms are held together by strong **covalent** bonds.

Word and Symbol Equations

Symbol equations can be used to describe what happens during a chemical reaction.

When magnesium burns in air the magnesium metal reacts with the non-metal atoms in oxygen molecules to form the ionic compound magnesium oxide. This reaction can be shown in a word equation:

Magnesium + Oxygen → Magnesium Oxide

or by the symbol equation:

$2Mg + O_2 \rightarrow 2MgO$

Atoms are not created or destroyed during a chemical reaction: the atoms are just rearranged.

This means that the total mass of the **reactants** is the same as the total mass of the **products**.

Ionic Compounds

Ionic compounds are formed when a metal reacts with a non-metal. When metal atoms react they lose negatively charged electrons to become positively charged ions (or **cations**). When non-metal atoms react they gain negatively charged electrons to become negatively charged ions (or **anions**).

Ionic compounds

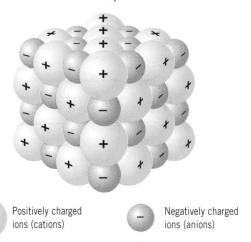

+ Positively charged ions (cations)
− Negatively charged ions (anions)

Build Your Understanding

There is no overall charge on ionic compounds so you can use the charge on the ions to work out the formula of the ionic compound.

Metal Ions	**Non-metal Ions**
Sodium, Na^+	Bromide, Br^-
Potassium, K^+	Chloride, Cl^-

The compound sodium chloride contains sodium, Na^+, and chloride, Cl^-, ions. For every one sodium ion one chloride ion is required. The overall formula for the compound is NaCl.

❓ Test Yourself

1. How can atoms join together?
2. Give the name of the elements with the symbols Na and Cr.
3. A water molecule has the formula H_2O. Explain what this formula tells us.
4. Sodium nitrate has the formula $NaNO_3$. Explain what this formula tells us.

⭐ Stretch Yourself

1. Give the formula for the following compounds:
   a) Potassium chloride.
   b) Sodium bromide.

Balancing Equations

Conservation of Mass

Symbol equations show the type and ratio of the atoms involved in a reaction. The reactants are placed on the left-hand side of the equation. The products are placed on the right-hand side of the equation.

Overall, mass is **conserved** because atoms are never made or destroyed during chemical reactions. This means that there must always be the same number of each type of atom on both sides of the equation.

Balancing the Equation

Hydrogen burns in air to produce water vapour. This can be shown using a word equation:

Hydrogen + Oxygen → Water

The word equation is useful, but it doesn't give the ratio of hydrogen and oxygen molecules (small groups of atoms joined by covalent bonds – where atoms share pairs of electrons) involved. Balanced symbol equations show this extra information. First, replace the words with symbols.

Hydrogen and oxygen both exist as molecules:

$H_2 + O_2 \rightarrow H_2O$

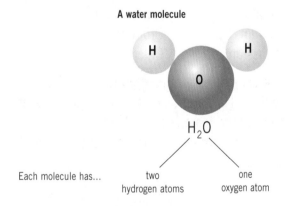

A water molecule

H_2O

Each molecule has... two hydrogen atoms, one oxygen atom

The formulae are all correct, but the equation does not balance because there are different numbers of atoms on each side of the equation. The formulae cannot be changed, but numbers can be added in front of the formulae to balance the equation.

The equation shows that there are two oxygen atoms on the left-hand side of the equation, but only one oxygen atom on the right-hand side.

A number 2 is therefore placed in front of the H_2O:

$H_2 + O_2 \rightarrow 2H_2O$

Now the oxygen atoms are balanced: there are the same number of oxygen atoms on both sides of the equation. However, the hydrogen atoms are no longer balanced. There are two hydrogen atoms on the left-hand side and four hydrogen atoms on the right-hand side. So in front of the H_2 a 2 is placed:

$2H_2 + O_2 \rightarrow 2H_2O$

The equation is now balanced.

When balancing equations, always check that the formulae you have written down are correct.

Some equations involve formulae that contain brackets, for example calcium hydroxide, $Ca(OH)_2$. This means that calcium hydroxide contains calcium, oxygen and hydrogen atoms in the ratio 1 : 2 : 2. These equations can be balanced normally.

✓ Maximise Your Marks

To get a top mark, you need to be able to balance equations. This skill just needs a little practice. Deal with each type of atom in turn until everything balances.

Remember to write any subscripts below the line: H_2O is correct while H²O and H2O are wrong.

State Symbols

State symbols can be added to an equation to show extra information. They show what state the reactants and products are in.

The symbols are:
- (s) for solid
- (l) for liquid
- (g) for gases
- (aq) for aqueous.

Aqueous comes from the Latin *aqua* meaning water. Aqueous means dissolved in water.

Magnesium metal can be burned in air to produce magnesium oxide. Magnesium and magnesium oxide are both solids. The part of the air that reacts when things are burned is oxygen, which is a gas:

Magnesium + Oxygen → Magnesium Oxide

$2Mg(s) + O_2(g) \rightarrow 2MgO(s)$

Precipitation Reactions

A precipitate is a solid formed when two solutions react together.

Some **insoluble salts** can be made from the reaction between two solutions.

Barium sulfate is an insoluble salt. It can be made by the reaction between solutions of barium chloride and sodium sulfate:

Barium Chloride + Sodium Sulfate → Barium Sulfate + Sodium Chloride

$BaCl_2(aq) + Na_2SO_4(aq) \rightarrow BaSO_4(s) + 2NaCl(aq)$

Build Your Understanding

The insoluble salt barium sulfate can be filtered off, washed and dried.

Overall, the two original salts, barium chloride and sodium sulfate, have swapped partners. This can be described as a **double decomposition reaction**. Barium chloride solution can be used to test whether a solution contains sulfate ions. If sulfate ions are present, a white precipitate of barium sulfate will be seen.

The chloride ions and sodium ions are **spectator ions**. They are present, but they are not involved in the reaction. The ionic equation for the reaction is:

$Ba^{2+}(aq) + SO_4^{2-}(aq) \rightarrow BaSO_4(s)$

Precipitation reactions are very fast. When the reactant solutions are mixed, the reacting ions collide together very quickly and react together to form the insoluble solid.

Barium sulfate is used in medicine as a **barium meal**. The patient is given the insoluble salt and then X-rayed. The barium sulfate is opaque to X-rays so doctors can detect digestive problems without having to carry out an operation. Although barium salts are toxic, barium sulfate is so insoluble that very little dissolves and passes into the bloodstream of the patient.

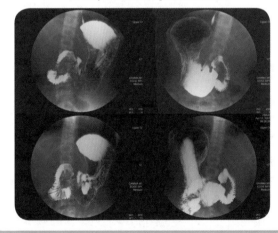

An X-ray taken following a barium meal

❓ Test Yourself

1. Why must there be the same number of each type of atom on both sides of an equation?
2. Balance the equation $Na + Cl_2 \rightarrow NaCl$.
3. Balance the equation $H_2 + Cl_2 \rightarrow HCl$.

⭐ Stretch Yourself

1. Explain why precipitation reactions happen very quickly.

4. What does the state symbol (l) indicate?

Ionic and Covalent Bonding

Types of Bonding

Compounds are made when atoms of two or more elements are chemically combined.

Ionic bonding involves the transfer of electrons in the outermost shell of atoms. This forms **ions** with opposite charges, which then attract each other. Ions are atoms or groups of atoms with a charge.

Covalent bonding involves the sharing of pairs electrons. The attraction between shared pairs of electrons holds the atoms together.

Ionic Bonding

Atoms react to get a full outer shell of electrons. Ionic bonding involves the transfer of electrons from one atom to another.

Metal atoms in groups 1 and 2, such as sodium or calcium, lose electrons to get a full outer shell of electrons. Overall, they become positively charged (electrons have a negative charge). Non-metal atoms in groups 6 and 7, such as oxygen or chlorine, gain electrons to get a full outer shell. They become negatively charged.

An ion is an atom, or a group of atoms, with a charge. An atom, or group of atoms, becomes an ion by gaining or losing electrons. The positive and negative ions formed have the same electronic structure as a noble gas atom.

Build Your Understanding

Ionic Compounds

Sodium reacts with chlorine to make sodium chloride:

Sodium + Chlorine → Sodium Chloride

- Each sodium atom transfers one electron from its outer shell to a chlorine atom.
- The sodium atom has lost a negatively charged electron, so it now has a 1+ charge and is called a sodium ion.
- The chlorine atom has gained an electron so it has a 1– charge and is now a chloride ion.
- Both the sodium ion and chlorine ion have a full outer shell.
- The attraction between these two oppositely charged ions is called an ionic bond, which holds the compound together.

In dot and cross diagrams, the electrons drawn as dots, and the electrons drawn as crosses are identical. They are drawn like this so it easier to see what happens when the electrons move.

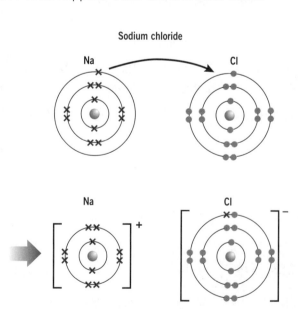

Build Your Understanding (cont.)

Magnesium Oxide

Magnesium reacts with oxygen to make magnesium oxide:

Magnesium + Oxygen → Magnesium Oxide

- The magnesium atom transfers two electrons from its outer shell to the oxygen atom.
- The magnesium atom has lost two electrons so has a 2+ charge. It is now a magnesium ion.
- The oxygen atom has gained two electrons so has a 2− charge. It is now an oxide ion.
- Both the magnesium and oxygen atoms have a full outer shell.
- The attraction between these two oppositely charged ions is called an ionic bond, which holds the compound together.

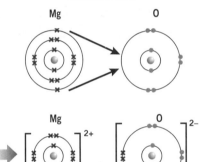

Magnesium oxide has a higher melting point than sodium chloride because magnesium oxide contains smaller ions that have higher charges, so the attraction between these ions is stronger.

Make sure you can apply these ideas to other examples.

Covalent Bonding

Covalent bonding occurs between atoms of non-metal elements. The atoms share pairs of electrons so that all the atoms gain a full outer shell of electrons.

There is an **electrostatic attraction** between the nuclei of the atoms and the bonding electrons.

Hydrogen, H_2

Both the hydrogen atoms have just one electron. Both atoms can get a full outer shell by sharing their electrons to form a single covalent bond.

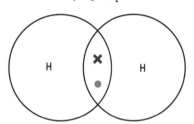

Oxygen, O_2

Both oxygen atoms have six outer electrons so they share two pairs of electrons to form a double covalent bond.

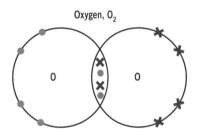

Make sure you can apply these ideas to other examples.

Test Yourself

1. What are ions?
2. What happens to electrons during the formation of ionic bonds?
3. What holds the atoms together in covalent molecules?
4. What charge does a sodium ion have?

Stretch Yourself

1. Name and describe the type of bonding that you would expect to find in these substances:
   a) Oxygen, O_2.
   b) Sodium chloride, NaCl.

Ionic and Covalent Structures

Ionic Bonding

Ionic bonding occurs between metal and non-metal atoms. It involves the transfer of electrons and the formation of ions. Sodium chloride and magnesium oxide are examples of ionic compounds.

Build Your Understanding

Ionic compounds are held together by the strong forces of attraction between oppositely charged ions (electrostatic attraction).

Ionic compounds have a regular structure.

The strong forces of attraction between oppositely charged ions work in all directions and this means that ionic compounds have high melting and boiling points.

When dissolved in water, ionic compounds form solutions in which the ions can move so these solutions can conduct electricity.

Similarly, if ionic compounds are heated up so that they melt, the ions can move. Molten ionic compounds can also conduct electricity.

Simple Covalent Structures

Covalent bonding occurs between non-metal atoms. It involves the sharing of pairs of electrons. Examples of simple covalent structures include chlorine and oxygen.

These molecules are formed from small numbers of atoms. The low boiling points of simple molecules is the result of the weak forces of attraction between molecules.

Oxygen is an example of a simple covalent structure

Properties of Simple Covalent Structures

In simple covalent structures there are very strong covalent bonds between the atoms in each molecule, but very weak forces of attraction between these molecules.

This means that molecular compounds have low melting and boiling points; most are gases or liquids at room temperature. Simple molecular compounds do not conduct electricity because, unlike ions, the molecules do not have an overall electrical charge. They tend to be insoluble in water (although they may dissolve in other solvents).

💡 Boost Your Memory

To help you learn the facts you need for the exam, make a table to compare the features of ionic and covalent structures.

Giant Covalent Structures

Examples of giant covalent (macromolecular) structures include diamond, graphite and silicon dioxide.

These structures are formed from a large number of atoms.

Properties of Giant Covalent Structures

The atoms in giant covalent structures are held together by **strong covalent bonds**. This means that these substances have high melting and boiling points. They are solids at room temperature. Like simple covalent molecules, giant covalent substances do not conduct electricity (except graphite) as they do not contain ions. They are also insoluble in water.

Diamond

Diamond is an example of a giant covalent substance. Like other gemstones, it is prized for its rarity and its pleasing appearance: it is lustrous, colourless and transparent. Diamond is also very hard.

High quality diamonds are used to make jewellery. Other diamonds are used in industry in a variety of applications.

The hardness and high melting point of diamond makes it very suitable for cutting tools.

The special properties of diamond are the result of its **structure**.

Diamond Bonding
Each carbon atom is bonded to four other carbon atoms by strong covalent bonds. It takes a lot of energy to break these strong bonds.
Diamond is a very poor electrical conductor because it does not have any free electrons.

Silicon dioxide, SiO_2, is found in the mineral quartz. It is very hard, which makes it resistant to weathering and it is the main constituent in sandstone. It is sometimes called silica.

Silicon dioxide has a similar structure to diamond and each silicon atom is attached to oxygen atoms by covalent bonds. Silicon dioxide does not conduct electricity because all the electrons are held in strong covalent bonds and cannot move.

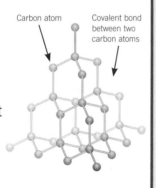

Graphite

Allotropes are different forms of the same element in the same physical state. Diamond, graphite and fullerene are all allotropes of carbon.

Graphite is lustrous, black and opaque. It is used to make pencil 'leads' because the layers slide over each other easily, so when a pencil is rubbed on paper a black mark is left.

Graphite is also used in lubricants because it is slippery and allows surfaces to pass over each other more easily.

Graphite Bonding
In graphite, each carbon atom forms strong covalent bonds with three other carbon atoms in the same layer. However, the bonding between layers is much weaker so the layers can pass over each other quite easily, which is why graphite is soft and feels greasy.

If a potential difference is applied across graphite, the electrons in the weak bonds between layers move and so conduct electricity.

Carbon in the form of graphite is the only non-metal element that conducts electricity. It also has a very high melting point because a lot of energy is required to break these strong covalent bonds. These properties make it a suitable material from which to make electrodes.

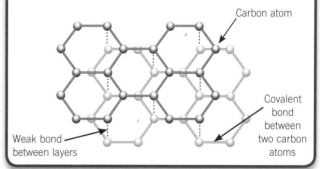

? Test Yourself

1. What type of structure is:
   a) magnesium oxide?
   b) graphite?
   c) diamond?

★ Stretch Yourself

1. Sodium chloride is an ionic compound. It does not conduct electricity when solid, but does when it is dissolved in water. Explain these observations in terms of the particles involved.

Group 7

The Halogens

The elements in group 7 are known as the **halogens**. The atoms of group 7 elements all have seven electrons in their outermost shell. When halogen atoms react they gain an electron to form **halide ions**.

Group 7 elements have similar properties because they all have similar electron structures. Halogens react with metal atoms to form ionic compounds, for example chlorine reacts with potassium to form potassium chloride. Group 7 atoms form molecules in which two atoms are joined together. These are called **diatomic** molecules.

The halogen family includes fluorine, chlorine, bromine and iodine. Halogens have coloured vapours. The colour gets darker further down the group.

Build Your Understanding

Down the group, the melting points and boiling points of the halogens increase, so fluorine and chlorine are gases at room temperature while bromine is a liquid and iodine is a solid.

Halogens react with hydrogen to form hydrogen halides, for example chlorine reacts with hydrogen to form hydrogen chloride. Hydrogen halides dissolve in water to form acidic solutions.

The atoms get larger and have more electrons further down the group. This means that the strength of the attraction between molecules increases.

As the forces of attraction between molecules get stronger down the group, it takes more energy to overcome these forces so the halogens will melt and boil at higher temperatures.

Physical Properties and Uses

Fluorine is a very poisonous gas that should only be used in a fume cupboard. Fluorine is a diatomic molecule with the formula F_2. The gas has a pale yellow colour.

Sodium fluoride is added to toothpastes and to some water supplies to help prevent tooth decay. Scientists carried out large studies to prove that adding fluoride compounds was effective at protecting teeth, but some people are concerned over the lack of choice those living in affected areas now have.

Chlorine is a poisonous gas that should only be used in a fume cupboard. Chlorine is a diatomic molecule with the formula Cl_2. The gas has a pale green colour.

Chlorine kills bacteria and is used in water purification. It is also used to make plastics and pesticides and in bleaching.

In the past, chlorine and iodine were extracted from compounds found in seawater. However, it is no longer economically worthwhile to extract iodine in this way.

Bromine is a poisonous, dense liquid. It has a brown colour.

Iodine exists as a black crystalline solid. Solid iodine is brittle, crumbly and is a poor electrical and thermal conductor. Iodine forms a purple vapour when warmed.

Iodine solution can be used as an antiseptic to sterilise wounds because it kills bacteria and can be used to test for the presence of starch. When iodine solution is placed on a material that contains starch it turns blue/black.

Astatine is found just below iodine in the periodic table. We can use the physical properties of the other halogens to predict the properties of astatine. It will be a dark coloured solid at room temperature.

Why Chlorine Reacts More Vigorously than Bromine

Reactivity decreases down the group: as an atom reacts to form an ion, the new electron is being placed into a shell further away from the nucleus. So, down the group, it is harder for atoms to gain an electron. There are also more shells of electrons shielding the new electron from the nucleus. This also makes it harder for atoms to gain a new electron further down the group. This pattern is clearly shown by the reaction between the halogens and iron wool to form iron halides.

Halogen Used	Observations
Chlorine	The iron glows very brightly. A brown smoke is given off and a brown solid is formed.
Bromine	The iron glows. Brown smoke is given off and a brown solid is formed.

💡 Boost Your Memory

Chlorine is more reactive than bromine and iodine.

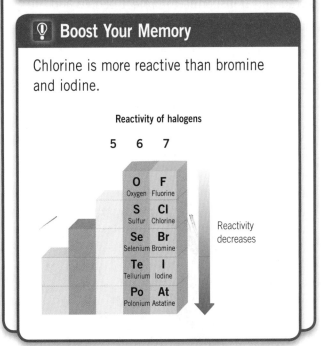

Reactivity of halogens

Displacement Reactions Involving Halogens

Reactivity decreases down group 7. The most reactive halogen is fluorine, followed by chlorine, then bromine, then iodine.

A more reactive halogen will **displace** (that is, take the place of) a less reactive halogen from an aqueous solution of its salt. So, chlorine could displace bromine and iodine.

However, while bromine could displace iodine it could not displace chlorine.

Chlorine will displace iodine from a solution of potassium iodide:

Chlorine + Potassium Iodide → Iodine + Potassium Chloride

$$Cl_2 + 2KI \rightarrow I_2 + 2KCl$$

Build Your Understanding

Chlorine reacts with potassium to make potassium chloride:

Potassium + Chloride → Potassium Chloride

$$2K + Cl_2 \rightarrow 2KCl$$

When they react, a halogen atom gains an electron to form an ion with a 1– charge:

$$Cl + e^- \rightarrow Cl^-$$

A reduction reaction has taken place. The halogen atom has gained an electron so it is reduced.

❓ Test Yourself

1. What is the name given to group 7 of the periodic table?
2. How is a halide ion formed?
3. What type of compound is formed when a metal reacts with a halogen?

⭐ Stretch Yourself

1. Chlorine gas is passed through an aqueous solution of potassium iodide. Write word and symbol equations to sum up this reaction.

New Chemicals and Materials

Making New Chemicals

The manufacture of useful chemicals involves many stages. **Raw materials** need to be selected and prepared, and then the new chemicals have to be made in a process known as **synthesis**.

Next, the useful products have to be *separated* from **by-products** and waste, each of which must also be dealt with. Finally, the **purity** of the product must be checked.

Build Your Understanding

Some chemicals are made in batch processes which are used to make relatively small amounts of special chemicals such as medicines. The chemicals are made when they are needed, rather than all the time.

Continuous processes are used to make chemicals that are needed in large amounts, such as sulfuric acid or ammonia. These chemicals are made all the time. Raw materials are continuously added and the new products are removed.

Some chemicals, such as ammonia, sulfuric acid, sodium hydroxide and phosphoric acid, are made in bulk (on a large scale). Other chemicals, such as medicines, food additives and fragrances, are described as being made on a fine scale (a small scale).

Governments regulate how chemicals are made, stored and transported to protect people and the environment from accidental damage.

Buckminster Fullerene, C_{60}

The element carbon exists in three forms or allotropes:
- graphite
- diamond
- fullerenes.

Fullerenes are structures made when carbon atoms join together to form tubes, balls or cages, which are held together by strong covalent bonds. The most symmetrical and most stable example is buckminster fullerene. This is a new material scientists have discovered, which consists of 60 carbon atoms joined together in a series of hexagons and pentagons, much like a leather football.

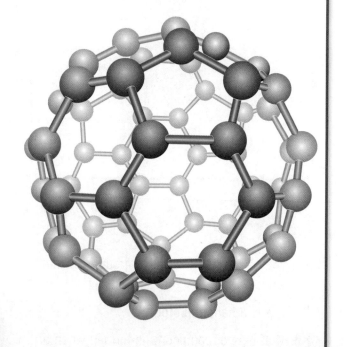

Structure of buckminster fullerene

✓ Maximise Your Marks

Learn the formula for buckminster fullerene: C_{60}.

Nanoparticles

Nanoscience is the study of extremely small pieces of material called **nanoparticles**. Scientists are currently researching the properties of new nanoparticles.

These are substances that contain just a few hundred atoms and vary in size from 1 nm (nanometres) to 100 nm (human hair has a width of about 100 000 nm). Nanoparticles occur in nature, for example in sea spray. They can also be made accidentally, for example when fuels are burned.

Nanomaterials have unique properties because of the very precise way in which the atoms are arranged. Scientists have found that many materials behave differently on such a small scale.

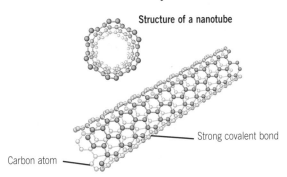

Structure of a nanotube — Carbon atom, Strong covalent bond

Lightweight Materials

Scientists are using nanoparticles to develop lightweight materials. These materials are incredibly hard and strong because of the precise way that the atoms are arranged. One day these materials could be used to build planes.

✓ Maximise Your Marks

In nanomaterials, the atoms themselves are not smaller. When you answer exam questions, make sure you do not infer that the atoms have changed size.

Other Uses of Nanoparticles

Nanoparticles have a very high surface area to volume ratio. Scientists hope that this will allow them to use nanoparticles in exciting ways such as:
- In new computers.
- In sunscreens and deodorants.
- In drug delivery systems.
- As better catalysts.

Catalysts are substances that speed up the rate of a chemical reaction, but are not themselves used up. Reactions take place at the surface of the catalyst. The larger the surface area of the catalyst, the more changes can take place at once and the better the catalyst performs.

Scientists are also keen to explore the use of nanoparticles as sensors to detect biological or chemical agents at very low levels. They may also be used to make battery electrodes for electric vehicles or solar cells.

Nanoscale silver particles have antibacterial, antiviral and antifungal properties. These tiny pieces of silver are incorporated into materials to make clothes and medical dressings stay fresh for longer.

There has recently been a great deal of media interest in the development and applications of new nanoparticles. Some scientists are concerned that certain nanoparticles could be dangerous to people because their exceptionally small size may mean they are able to pass into the body in previously unimaginable ways, and could go on to cause health problems.

💡 Boost Your Memory

Try producing a set of revision cards to learn the important ideas in this topic.

❓ Test Yourself

1. How big are nanoparticles?
2. What is the formula of buckminster fullerene?
3. Where are nanoparticles found in nature?

⭐ Stretch Yourself

1. Describe what happens during a continuous process.

Plastics and Perfumes

Polymerisation

Plastics are **synthetic** (manufactured) **polymers**. **Natural polymers** include cotton, wood, leather, silk and wool. In polymers, lots of small molecules are joined together to make one big molecule.

The simplest alkene, ethene, can be formed by the cracking of large hydrocarbon molecules. If ethene is heated under pressure in the presence of a **catalyst**, many ethene molecules can join together to form a compound composed of large molecules called poly(ethene) or polythene. The diagram shows how a large number of ethene molecules join together to form polythene.

$$n \begin{array}{c} H \;\; H \\ C=C \\ H \;\; H \end{array} \rightarrow \left(\begin{array}{c} H \;\; H \\ C-C \\ H \;\; H \end{array} \right)_n$$

The 'n' at the start of the equation and the section of the polymer surrounded by brackets represents the number of molecules involved. The brackets are used because it would be impractical to write out the complete structure. The brackets surround a representative unit that is then repeated through the whole polymer.

The small starting molecules, in this case the ethene molecules, are called **monomers**. The ethene molecules join together to form long chain molecules called polymers. A polymer is made from lots of monomer units.

The polythene diagram is an example of an **addition polymerisation** reaction. The ethene molecules have simply joined together.

Other Polymers

Polymerisation reactions may involve other monomer units. The exact properties of the polymer formed depend upon:
- The monomers involved.
- The conditions under which it was made.
- The length of the polymer chains.

Low density polythene and high density polythene have very different properties and uses because they are produced using different catalysts and different conditions. A plastic's properties are also affected by the **crystallinity** of its structure; the more crystalline a plastic is the more brittle it will be.

A crystalline polymer

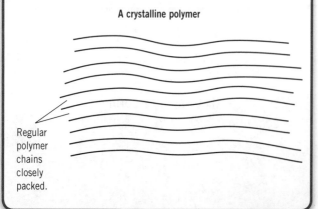

Regular polymer chains closely packed.

Build Your Understanding

Polypropene is made by an addition polymerisation reaction between many propene molecules.

$$n \begin{array}{c} CH_3 \;\; H \\ C=C \\ H \;\; H \end{array} \rightarrow \left(\begin{array}{c} CH_3 \;\; H \\ C-C \\ H \;\; H \end{array} \right)_n$$

Polyvinyl chloride (PVC) is made by an addition polymerisation reaction between many chloroethene molecules. Chloroethene used to be called vinyl chloride.

$$n \begin{array}{c} Cl \;\; H \\ C=C \\ H \;\; H \end{array} \rightarrow \left(\begin{array}{c} Cl \;\; H \\ C-C \\ H \;\; H \end{array} \right)_n$$

Polytetrafluoroethene (PTFE or Teflon™) is made by an addition polymerisation reaction between many tetrafluoroethene molecules.

$$n \begin{array}{c} F \;\; F \\ C=C \\ F \;\; F \end{array} \rightarrow \left(\begin{array}{c} F \;\; F \\ C-C \\ F \;\; F \end{array} \right)_n$$

PTFE is known as 'Teflon'. Surfaces coated in Teflon™ have low friction. It is used to coat some frying pans and saucepans.

Thermoplastics and Thermosetting Plastics

Thermoplastics consist of long polymer chains with few cross-links. When heated, these chains untangle and the material softens. It can then be reshaped. On cooling, the material becomes solid and stiff again. Thermoplastics can be heated and reshaped many times. Polythene is a thermoplastic. Thermoplastics can be stretched easily.

Thermosetting plastics consist of long, heavily cross-linked polymer chains. When they are first made these thermosetting plastics are soft and can be shaped. Once they have set, however, they become solid and stiff. They do not soften again, even if they are heated to very high temperatures, and so they cannot be reshaped. Thermosetting plastics, such as melamine, are rigid and cannot be stretched.

Esters

Esters are a family of organic compounds formed when alcohols react with carboxylic acids. Esters have pleasant fruity smells and flavours. Esters are also used as solvents and as **plasticisers**.

The ester ethyl ethanoate is formed when ethanol is reacted with ethanoic acid using an acid **catalyst**.

Esters are very useful chemicals. Esters are described as being **volatile** because they **evaporate** easily.

They are used as cheap alternatives to naturally occurring compounds in perfumes and body sprays, and as flavourings in foods such as yoghurts.

Esters can be used as flavourings

Perfumes

Traditional perfumes contain plant and animal extracts, such as rose and musk.

Today, cheaper **synthetic** fragrances, including esters, are often used. These are alternatives to materials made from living things.

A good perfume should:
- Evaporate easily from the skin so that the particles can be smelt.
- Be **non-toxic** and not react with water (so it does not react with sweat).
- Not irritate the skin (so it does not damage the skin).
- Be **insoluble** in water (so it is not washed off easily).

Build Your Understanding

Perfumes evaporate because, although there are strong forces of attraction *within* perfume particles, there are weaker forces of attraction *between* perfume particles. When the perfume is put on the skin, some of the particles gain enough energy to evaporate. The perfume particles can then travel through the air and be smelt.

Test Yourself

1. What type of reaction is involved in the formation of polythene?
2. What is the name given to the small units that join together to form a polymer?
3. Why are synthetic particles used in some perfumes?

Stretch Yourself

1. Draw a tetrafluoroethene molecule.
2. Is tetrafluoroethene saturated or unsaturated? Explain your answer.
3. Why must perfumes evaporate easily?

Analysis

Modern Methods of Analysis

Compared to the more traditional laboratory methods, modern instrumental methods of analysing chemicals are faster, more sensitive, more accurate and require smaller sample sizes.

Qualitative analysis allows scientists to identify which components are present; **quantitative** analysis allows them to decide how much of each component is present.

It is important that scientists test a sample that is representative of the whole material being tested and that any sample is collected, stored and prepared carefully to prevent contamination, which would invalidate the results.

Many analytical techniques use samples that have been dissolved to form solutions.

Chromatography

Chromatography is used to **separate** the components of a mixture. It is used to analyse colouring agents in foods, flavourings and drugs. The technique is used in the food industry and by forensic scientists.

A small spot of the substance being analysed is placed on a pencil line towards the bottom of the chromatography paper. The chromatography paper is then placed end down into a beaker containing a small amount of the solvent being used. The solvent moves up the paper, carrying the soluble components. When the solvent front nearly reaches the top of the paper the paper is removed and the solvent is allowed to evaporate:
- Aqueous solvents contain water and are useful for many ionic compounds.
- Non-aqueous solvents do not contain water.

The more soluble a component is in the moving solvent the further it will move up the chromatography paper.

Build Your Understanding

The Rf value is used to compare the distance the component has moved compared to the distance the solvent front has moved.

$$Rf = \frac{\text{distance moved by the component}}{\text{distance moved by the solvent front}}$$

The solvent front has moved 6.0 cm.

The yellow spot has moved 3.0 cm.

The Rf value for the yellow spot = $\frac{3.0 \text{ cm}}{6.0 \text{ cm}}$ = 0.50

If the same solvent and the same conditions are used, the Rf value would be the same for a given component. If the Rf value for an unknown compound is determined it can be identified by comparing this value with the Rf values for known compounds. Unfortunately, similar compounds often have quite similar Rf values.

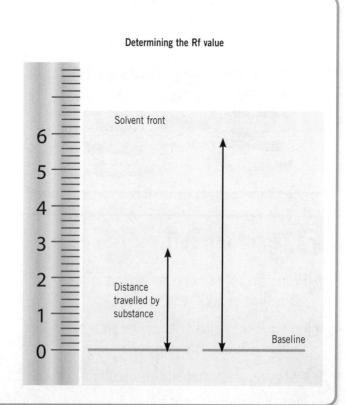

Determining the Rf value

Mass Spectrometry

Mass spectrometry is used to find the accurate relative atomic or formula mass of a compound. Gas chromatography can be linked to mass spectrometry in a technique known as **GC-MS**. As the sample leaves the gas chromatogram, it is fed into the mass spectrum and the relative molecular mass of each substance can be identified.

The heaviest peak in the sample is known as the molecular ion and can be used to identify the molecular mass of the compound.

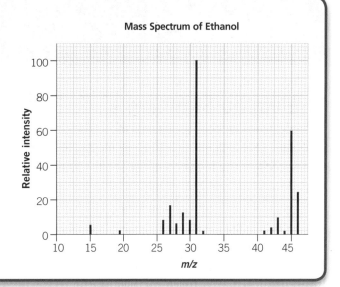

Gas Chromatography

Gas chromatography is used to identify organic compounds with low boiling points. The retention time is the time that it takes a component to pass through the column of the gas chromatogram.

Different components take different times to move through this column and so have different retention times. Unknown compounds can be identified by comparing their retention times with the retention times for known compounds. The areas under the peaks in the chromatogram are proportional to the amount of each compound in the sample. The number of peaks in the chromatogram shows the number of components in the sample.

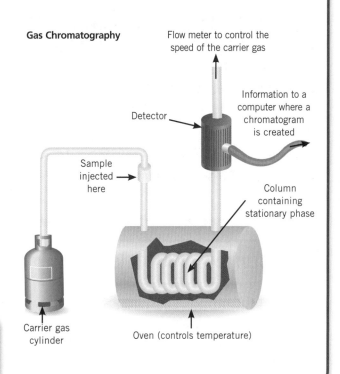

✓ Maximise Your Marks

Learn how gas chromatography and thin layer chromatography work.

? Test Yourself

1. What are the advantages of modern methods of instrumental analysis over traditional methods?
2. Why is chromatography useful?

★ Stretch Yourself

1. In separating a sample of food colour containing a blue and a green component, the chromatogram produced showed the solvent front moved 8.0 cm while the green component moved 6.0 cm and the blue component moved 5.0 cm. Calculate the Rf values for the green and blue components.

Metals

Metallic Structure

Metals have a giant structure. In metals, the electrons in the highest energy shells (outer electrons) are not bound to one atom but are **delocalised**, or free to move through the whole structure. This means that metals consist of positive metal ions surrounded by a 'sea' of negative electrons. **Metallic bonding** is the attraction between these positive ions and the negative electrons. This is an **electrostatic** attraction.

✓ Maximise Your Marks

Remember, the metallic bond is the electrostatic attraction between the positive metal ions and the delocalised electrons.

Remember to say that metals conduct electricity because the delocalised electrons can move. Do not talk about atoms or ions moving.

Metallic structure

Moving electrons can carry the electric charge or thermal (heat) energy.

Properties of Metals

Metallic bonding means that metals have several very useful properties:
- The free electrons mean that metals are **good electrical conductors**.
- The free electrons also mean that metals are **good thermal conductors**.
- Metals can be drawn into wires as the ions slide over each other.
- Metals can also be hammered into shape (they are malleable).
- Most metals have **high melting points** because lots of energy is needed to overcome the strong metallic bonds.

Non-metals are found on the right-hand side of the periodic table. They tend to be **poor electrical and thermal conductors**. Non-metals generally have **low melting points** and boiling points and are sometimes gases at room temperature.

Smart Alloys

Smart alloys are new materials with amazing properties.

One famous example of a smart alloy is **nitinol**. Nitinol is an alloy of nickel and titanium.

Some smart alloys have a **shape memory**. When a force is applied to a smart alloy it **stretches** or bends. When a smart alloy is heated up, however, it returns to its original shape.

Smart alloys appear to have a shape memory because they are able to exist in two solid forms. A temperature change of 10–20 degrees is enough to cause smart alloys to change forms.

Shape-memory polymers behave in a similar way, returning to their original shape when heated.

Smart Alloys (cont.)

At low temperatures, smart alloys exist in their low temperature form.

Low temperature form

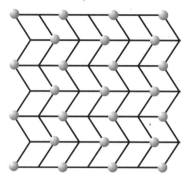

If a force is applied to the alloy it can be distorted to the low temperature, deformed form of the alloy.

Low temperature, deformed form of the alloy

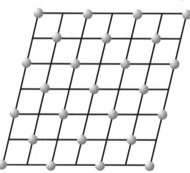

When the alloy is heated, it changes to the higher temperature form.

Higher temperature form

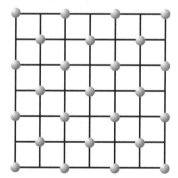

In shape-memory alloys, the low temperature form and the high temperature form are the same shape and size, so when they are heated smart alloys appear to have a shape memory. Nitinol is used in some dental braces.

Shape-memory alloys are also used in spectacle frames and as stents in damaged blood vessels.

Superconductors and Electrical Resistance

Some metals can behave as **superconductors** at very low temperatures. Metals can conduct electricity because they have delocalised electrons that can move. Metals normally have a **resistance** to the current that is flowing. Energy is lost as the current overcomes this resistance and the metal warms up.

Build Your Understanding

Superconductors are special because they have little or no resistance. The advantages of using superconductors include:
- If there is no resistance then no energy is lost when a current flows.
- As the resistance decreases the current can flow faster, so super-fast circuits can be developed.
- They can be used to make powerful electromagnets.

Despite these advantages, superconductors are not widely used because they only work below a critical temperature. Although this varies for different superconductors, the current critical temperatures are around −170 °C; until they work at room temperature, their use is likely to be limited.

❓ Test Yourself

1. What is metallic bonding?
2. Why are metals good electrical conductors?
3. Which metals are used to make nitinol?
4. Why do metals have high melting points?

⭐ Stretch Yourself

1. Give a use of nitinol.
2. Why is the use of superconductors limited at present?

Group 1

The Alkali Metals

The **elements** in group 1, on the far left-hand side of the periodic table, are known as the **alkali metals**. They are soft metals with **low melting points**.

They react with water to form hydroxide solutions. They react with halogens to form salts.

Examples of alkali metals are lithium (Li), sodium (Na) and francium (Fr).

Build Your Understanding

Rubidium and caesium belong to group 1, but are too **reactive** for use in schools. Further down the group alkali metals get more reactive; they react more vigorously with water. Alkali metals are so reactive that they must be stored under oil to prevent them reacting with moisture or oxygen. Alkali metals are shiny when freshly cut, but they tarnish quickly as they react with oxygen. Gloves and goggles should be worn when using alkali metals.

Reaction with Water

Alkali metals have **low densities**; lithium, sodium and potassium are all less dense than water. The alkali metals become denser down the group. When alkali metal atoms react they lose the single electron in their outermost shell to form ionic compounds in which the alkali metal ions have a 1+ charge.

For example:

$Na \rightarrow Na^+ + e^-$

The alkali metals react with water to form strongly **alkaline hydroxide** solutions and hydrogen gas:

Metal + Water → Metal Hydroxide + Hydrogen

Metal	Observations When Metal Reacts with Water	Equation
Lithium, Li	Metal floats on water. Some bubbles seen.	Lithium + Water → Lithium Hydroxide + Hydrogen
Sodium, Na	Metal forms a molten ball that moves around on the surface of the water. Many bubbles seen.	Sodium + Water → Sodium Hydroxide + Hydrogen
Potassium, K	The metal reacts even more vigorously than sodium (it can ignite). Lots of bubbles are seen and the hydrogen formed burns with a lilac flame.	Potassium + Water → Potassium Hydroxide + Hydrogen

✓ Maximise Your Marks

To get a top grade, you need to be able to write the symbol equations to sum up these three reactions:

$2Li + 2H_2O \rightarrow 2LiOH + H_2$

$2Na + 2H_2O \rightarrow 2NaOH + H_2$

$2K + 2H_2O \rightarrow 2KOH + H_2$

Why Group 1 Metals All React in a Similar Way

Alkali metals have just one electron in their outer shell and so have similar properties because they have similar electron structures.

Alkali metals react with non-metals to form ionic compounds. For example, sodium reacts with chlorine to form sodium chloride:

Sodium + Chlorine → Sodium Chloride

When sodium is burned, it reacts with oxygen to form sodium oxide:

Sodium + Oxygen → Sodium Oxide

When they react, an alkali metal atom loses its outer electron to form an ion with a 1+ charge:

$Na \rightarrow Na^+ + e^-$

The alkali metal atom has lost an electron so it is **oxidised**.

Alkali metals form solid white ionic compounds that dissolve to form colourless solutions.

✓ Maximise Your Marks

To get a top grade, you need to be able to write the symbol equations to sum up these reactions:

$2Na + Cl_2 \rightarrow 2NaCl$

$4Na + O_2 \rightarrow 2Na_2O$

Build Your Understanding

Down the group the outermost electron is further from the nucleus.

Further down the group there are more shells shielding the outer electron from the atom's nucleus, so it is easier for atoms to lose their outer electron.

Why Melting and Boiling Points Decrease down the Group

Melting and boiling points decrease down the group.

Alkali metals are held together by metallic bonding. Metallic bonding is the attraction between the positive metal ions and the 'sea' of negative electrons.

Forces of Attraction

The atoms get larger down the group. The strength of the metallic bonding decreases because the forces of attraction become weaker further down the group so it takes less energy to overcome these forces and therefore the melting points and boiling points will decrease.

❓ Test Yourself

1. Name the first three metals in group 1.
2. How many electrons are present in the outer shell of all group 1 metals?
3. Why do all the group 1 metals have similar properties?
4. What type of compounds do group 1 metals form?

★ Stretch Yourself

1. Potassium reacts with chlorine to produce the compound potassium chloride.
   a) Write a word and symbol equation to sum up this reaction.
   b) Is the potassium oxidised or reduced in this reaction? Explain your answer.

Aluminium and Transitions Metals

The Extraction of Aluminium

Aluminium is more reactive than carbon and so it is extracted from its ore, bauxite, using **electrolysis**, even though this is a very expensive method.

For electrolysis to occur, the aluminium ions and oxide ions in bauxite must be able to move. This means that the bauxite has to be either heated until it melts or dissolved in something.

Build Your Understanding

Bauxite has a very high melting point and heating the ore to this temperature is very expensive. Fortunately, another ore of aluminium, called cryolite, has a much lower melting point. First, the cryolite is heated up until it melts and then the bauxite is dissolved in the molten cryolite. Extracting aluminium from its ore requires a lot more energy than extracting iron from its ore.

Electrolysis

Aluminium can be extracted by electrolysis.
- By dissolving the aluminium oxide in cryolite, both the aluminium, Al^{3+}, ions and the oxide, O^{2-}, ions can move.
- During electrolysis, the aluminium, Al^{3+} ions are attracted to the negative electrode (the cathode) where they pick up electrons to form aluminium, Al, atoms. The aluminium metal collects at the bottom of the cell where it can be gathered:
Aluminium ions + electrons → aluminium atoms
- The oxide, O^{2-}, ions are attracted to the positive electrode (the anode) where they lose electrons to form oxygen molecules:
Oxide ions − electrons → oxygen molecules

- The oxygen that forms at the positive electrode readily reacts with the carbon, graphite, electrode to form carbon dioxide. The electrodes, therefore, must be replaced periodically. Extracting aluminium is expensive because lots of energy is required and because there are lots of stages in the process.

✓ Maximise Your Marks

To get a top grade, make sure you can write equations for the reactions taking place at the electrodes:

$Al^{3+} + 3e^- \rightarrow Al$

$2O^{2-} - 4e^- \rightarrow O_2$

Oxidation and Reduction

In the electrolysis of aluminium oxide:
- Aluminium ions are reduced to aluminium atoms.
- Oxide ions are oxidised to oxygen molecules.

Reduction reactions happen when a substance gains electrons. In this case, each aluminium ion gains three electrons to form an aluminium atom.

Oxidation reactions occur when a species loses electrons. In this case, two oxide ions both lose two electrons to form an oxygen molecule.

Reduction and oxidation reactions must always occur together and so are sometimes referred to as **redox** reactions.

💡 Boost Your Memory

Oxidation and reduction can be remembered using the mnemonic **OIL RIG**:

Oxidation **I**s **L**oss **R**eduction **I**s **G**ain (of electrons).

Properties of Transition Metals

Transition metals are found in the middle section of the periodic table. Copper, iron and nickel are examples of very useful transition metals. All transition metals have characteristic properties:
- High **melting points** (except for mercury, which is a liquid at room temperature).
- A high **density**.

They form **coloured compounds**:
- Copper compounds are blue or green.
- Iron (II) compounds are green.
- Iron (III) compounds are a 'foxy' red.

Reactions of Transition Metals

Transition metals are strong, tough, **good thermal and electrical conductors**, malleable and hard wearing. All transition metals are much less reactive than group 1 metals. They react much less vigorously with oxygen and water. Many transition metals can form ions with different charges. This makes transition metals useful catalysts for many reactions.

Transition metals are hard wearing

Purification of Copper

Copper must be purified before it can be used for some applications, such as high-specification wiring.

Copper is purified using electrolysis:
- During the electrolysis of copper, impure copper metal is used as the positive electrode where copper atoms give up electrons to form copper ions.
- As the positive electrode dissolves away, any impurities fall to the bottom of the cell to form a sludge.
- Copper ions in the solution are attracted towards the negative electrode where the copper ions gain electrons to form copper atoms.
- The positive electrode gets smaller while the negative electrode gets bigger. In addition, the negative electrode is covered in very pure copper.

Electrolysis of copper

Positive electrode — This electrode dissolves
Negative electrode — Pure copper forms here.
Cu^{2+}
Sludge formed from impurities
Copper sulfate solution

? Test Yourself

1. What is the name of the method used to extract aluminium from its ore?
2. In which section of the periodic table are the transition metals found?
3. If you wanted to coat a metal object with copper, which electrode should you make it?

★ Stretch Yourself

1. In the extraction of aluminium, why is bauxite dissolved in molten cryolite?
1. Why are many transition metal compounds useful catalysts?

Chemical Tests

Flame Tests

Flame tests can be used to identify some metals present in salts. These elements give distinctive flame colours when heated because the light given out by a particular element gives a characteristic **line spectrum**. The technique of **spectroscopy** has been used by scientists to discover new elements, including caesium and rubidium:

- Clean a flame test wire by placing it into the hottest part of a blue Bunsen flame.
- Dip the end of the wire into water and then into the salt sample.
- Hold the salt in the hottest part of the flame and observe the colour seen.

Metal Ion Present	Colour in Flame Test
Lithium	Crimson
Sodium	Yellow/orange
Potassium	Lilac
Calcium	Red
Barium	Light green
Copper	Blue/green

✓ Maximise Your Marks

For flame tests, give the name of the metal ion responsible for a colour, not the name of a whole compound.

Hazard Symbols

Hazard symbols are a very effective way of alerting people to the dangers associated with different chemicals.

Toxic These substances can kill. They can act when you swallow them, breathe them in or absorb them through your skin. *Example: chlorine gas.*		**Corrosive** These substances attack other materials and living tissue, including eyes and skin. *Example: concentrated sulfuric acid.*	
Oxidising These substances provide oxygen, which allows other substances to burn more fiercely. *Example: hydrogen peroxide.*		**Irritant** These substances are not corrosive but they can cause blistering of the skin. *Example: calcium chloride.*	
Harmful These substances are similar to toxic substances but they are less dangerous. *Example: lead oxide.*		**Explosive** These substances are explosive. *Example: urea nitrate.*	
Highly Flammable These substances will catch fire easily. They pose a serious fire risk. *Example: hydrogen.*			

✓ Maximise Your Marks

Questions are often asked about hazard symbols. Make sure you can identify what each symbol shows and explain what it means.

Formula of Ionic Compounds

Metal Ions	Non-metal Ions
Sodium, Na^+	Oxide, O^{2-}
Magnesium, Mg^{2+}	Chloride, Cl^-
Calcium, Ca^{2+}	Bromide, Br^-
Potassium, K^+	Hydroxide, OH^-
Iron (II), Fe^{2+}	Nitrate, NO_3^-
Iron (III), Fe^{3+}	Carbonate, CO_3^{2-}
Copper (II), Cu^{2+}	Sulfate, SO_4^{2-}

The compound magnesium oxide contains magnesium, Mg^{2+}, and oxide, O^{2-}, ions. For every one magnesium ion, one oxide ion is required.

The overall formula for the compound is MgO.

Testing for Sulfate Ions

To test for the presence of sulfate ions in solution:
- Add dilute hydrochloric acid.
- Then add barium chloride solution.

A white precipitate of barium sulfate shows that sulfate ions are present in the original solution:

Barium + Sodium → Barium + Sodium
Chloride   Sulfate   Sulfate   Chloride

Symbol equation:

$$BaCl_2(aq) + Na_2SO_4(aq) \rightarrow BaSO_4(s) + 2NaCl(aq)$$

Ionic equation:

$$Ba^{2+}(aq) + SO_4^{2-}(aq) \rightarrow BaSO_4(s)$$

Testing for Halide Ions

It is important that a test for a particular ion gives a result that is **unique** to that ion for a positive identification to be made. These tests are very important and are used to check for the presence of chemicals in blood and to check the purity of drinking water.

Identifying Halide Ions			
Halide Ion	Test	Results	Ionic Equations
Chloride, Cl^-	Add dilute nitric acid then silver nitrate solution	Chloride ions give a white precipitate of silver chloride	$Ag^+(aq) + Cl^-(aq) \rightarrow AgCl(s)$
Bromide, Br^-	Add dilute nitric acid then silver nitrate solution	Bromide ions give a cream precipitate of silver bromide	$Ag^+(aq) + Br^-(aq) \rightarrow AgBr(s)$
Iodide, I^-	Add dilute nitric acid then silver nitrate solution	Iodide ions give a yellow precipitate of silver iodide	$Ag^+(aq) + I^-(aq) \rightarrow AgI(s)$

❓ Test Yourself

1. What is the formula of magnesium chloride?
2. What colour is the precipitate formed when dilute nitric acid then silver nitrate solution are added to a substance containing iodide ions?

⭐ Stretch Yourself

1. Describe how you would carry out a flame test.

Acids and Bases

Strong Acids

Acids and bases are chemical opposites. Some bases dissolve in water and are called alkalis.

Acidic solutions have a pH less than 7. Acidic compounds can be solids like citric acid and tartaric acid, liquids like sulfuric acid, nitric acid and ethanoic acid or gases like hydrogen chloride.

Some acids are described as strong. Examples of strong acids include hydrochloric acid, which is produced in the stomach and helps break down food and kills bacteria, sulfuric acid and nitric acid.

Strong acids are completely ionised in water. When hydrochloric acid is placed in water, every hydrogen chloride molecule splits up to form hydrogen ions and chloride ions:

Hydrochloric acid → Hydrogen + Chlorine

$HCl(g) + (aq) \rightarrow H^+(aq) + Cl^-(aq)$

Build Your Understanding

Other acids are described as weak acids. Examples of weak acids include ethanoic acid, citric acid and carbonic acid.

Weak acids do not completely ionise in water. When ethanoic acid is placed in water, only a small fraction of the ethanoic acid molecules split up to form hydrogen ions and ethanoate ions:

$CH_3COOH(l) \rightleftharpoons H^+(aq) + CH_3COO^-(aq)$

Notice that this reaction is reversible. Ethanoic acid reacts more slowly with metals and carbonates than a comparative amount of a strong acid like hydrochloric acid would do. This is because ethanoic acid produces fewer H^+ ions and so there are fewer collisions between reactant particles and H^+ ions.

A sample of a weak acid, like ethanoic acid, has a lower electrical conductivity than a sample of a strong acid, like hydrochloric acid, because hydrochloric acid is fully dissociated (that is, split up) in water and produces more H^+ ions.

However, both acids would produce the same volume of carbon dioxide if they were reacted with calcium carbonate or magnesium carbonate.

Electrolysis of both hydrochloric acid and ethanoic acid produces hydrogen gas at the negative electrode.

Weak acids, such as vinegar, are widely used as descalers since they remove limescale without damaging the surface of the object being cleaned.

Concentrated sulfuric acid is a dehydrating agent and can be used to remove water from sugar and from hydrated copper (II) sulfate.

✓ Maximise Your Marks

Acids are proton donors. Bases are proton acceptors.

Weak and Strong Alkalis

Traditional sources of alkalis included stale urine and burned wood. With industrialisation, the demand for alkalis grew, so shortages of alkalis soon developed.

Alkalis were used to neutralise acid soils, to produce the chemicals needed to bind dyes to cloth, to convert fats and oils into soap and to manufacture glass.

Weak and Strong Alkalis (cont.)

Early methods of manufacturing alkalis from limestone and salt produced a lot of **pollution**, including the acid gas hydrogen chloride and waste heaps that slowly released the toxic and unpleasant smelling gas hydrogen sulfide. **Oxidation** of hydrogen chloride forms chlorine gas.

Alkaline solutions have a pH more than 7. Some alkalis are described as **strong alkalis**. Examples of strong alkalis include sodium hydroxide and potassium hydroxide.

Strong alkalis are completely ionised in water. When sodium hydroxide is placed in water, it splits up to form sodium ions and hydroxide ions:

$NaOH(s) + (aq) \rightarrow Na^+(aq) + OH^-(aq)$

Other alkalis are described as weak. Ammonia is an example of a **weak alkali**.

Weak alkalis do not completely ionise in water. Ammonia produces hydroxide, OH^-, ions when it reacts with water:

Ammonia + Water ⇌ Ammonium + Hydroxide ion ion

$NH_3 + H_2O \rightleftharpoons NH_4^+ + OH^-$

Ammonium salts are useful **fertilisers**.

✓ Maximise Your Marks

Hydroxide ions have the formula OH^-. Remember to add the negative charge. Strong alkalis have a high pH and are fully ionised.

The pH Scale

The pH scale can be used to distinguish between weak and strong acids and alkalis. The pH scale measures the concentration of hydrogen ions. Neutral solutions have a pH of 7. Acidic solutions have a pH of less than 7.

The strongest acids have a pH of 1. Dilute solutions of weak acids have higher pH values than dilute solutions of strong acids. If water is added to an acid it becomes more dilute and less corrosive.

Alkaline solutions have a pH of more than 7. The strongest alkalis have a pH of 14. Many cleaning materials contain alkalis. If water is added to an alkali it becomes more dilute and less corrosive.

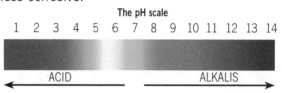

The pH scale

Indicators

Indicators can be used to show the pH of a solution. Indicators work by changing colour. They can show when exactly the right amount of acid and alkali have been added together.

Build Your Understanding

Red litmus turns blue in alkaline conditions; blue litmus turns red in acidic conditions.

❓ Test Yourself

1. Which ions are found in acidic solutions?
2. Which ions are found in alkaline solutions?
3. Name three strong acids and explain why they are described as 'strong'.

★ Stretch Yourself

1. A 1 g sample of calcium carbonate was placed in 50 cm³ of 1.0 mol dm⁻³ ethanoic acid. An identical sample of calcium carbonate was then placed in 50 cm³ of 1.0 mol dm⁻³ hydrochloric acid.
   a) In what ways would the reactions be the same?
   b) In what ways would the reactions be different?

Making Salts

Neutralisation Reactions

The reaction between an acid and a base is called **neutralisation**:
- Acidic solutions contain hydrogen, H^+, ions.
- Alkaline solutions contain hydroxide, OH^-, ions.

The reaction between an acid and an alkali can be shown in the word equation:

Acid + Alkali → Salt + Water

The ionic equation for all neutralisation reactions is:

$H^+(aq) + OH^-(aq) \rightarrow H_2O(l)$

The type of salt that is produced during the reaction depends on the acid and the alkali used. Indigestion medicines contain chemicals that react with, and neutralise, excess stomach acid.

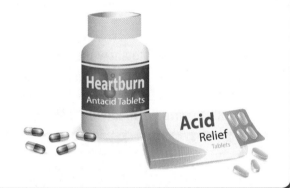

Naming Salts

Neutralising hydrochloric acid will produce **chloride salts**:

Hydrochloric + Sodium → Sodium + Water
Acid　　　　　Hydroxide　Chloride

Neutralising nitric acid will produce **nitrate salts**:

Nitric + Potassium → Potassium + Water
Acid　　Hydroxide　　Nitrate

Neutralising sulfuric acid will produce **sulfate salts**:

Sulfuric + Sodium → Sodium + Water
Acid　　　Hydroxide　Sulfate

Ammonia reacts with water to form a weak alkali. Ammonia solution can be neutralised with acids to form **ammonium salts**.

Build Your Understanding

Metal oxides are also bases. They can be reacted with acids to make salts and water:

Metal Oxide + Acid → Salt + Water

1. The metal oxide is added to an acid until no more will react.
2. Excess metal oxide is removed by filtering the solution.
3. The solution is evaporated to leave behind the salt.

Examples

Copper Oxide + Hydrochloric Acid → Copper Chloride + Water

$CuO + 2HCl \rightarrow CuCl_2 + H_2O$

Zinc Oxide + Sulfuric Acid → Zinc Sulfate + Water

$ZnO + H_2SO_4 \rightarrow ZnSO_4 + H_2O$

Copper Oxide + Nitric Acid → Copper Nitrate + Water

$CuO + 2HNO_3 \rightarrow Cu(NO_3)_2 + H_2O$

Build Your Understanding (cont.)

Copper + Sulfuric → Copper + Water
Oxide   Acid   Sulfate

$CuO + H_2SO_4 \rightarrow CuSO_4 + H_2O$

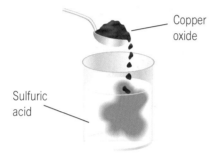

Add copper oxide to sulfuric acid

Filter to remove any unreacted copper oxide

Evaporate to leave behind blue crystals of the 'salt' copper sulfate

Making Salts from Metals

Some metals can be reacted with acids to form a salt and hydrogen.

Salts of very unreactive metals, such as copper, cannot be made in this way because these metals do not react with acids.

Salts of very reactive metals, such as sodium, cannot be made in this way because the reaction between the metal and acid is too vigorous to be carried out safely.

Precipitation Reactions

Some **insoluble** salts can be made from the reaction between two solutions. Barium sulfate is an insoluble salt. It can be made by the reaction between solutions of barium chloride and sodium sulfate:

Barium + Sodium → Barium + Sodium
Chloride Sulfate   Sulfate Chloride

Precipitation reactions can be used to remove unwanted ions from solutions. This technique is used to treat drinking water and effluent.

✓ Maximise Your Marks

To get a top grade, make sure you can write the symbol equation for this reaction:

$BaCl_2(aq) + Na_2SO_4(aq) \rightarrow BaSO_4(s) + 2NaCl(aq)$

❓ Test Yourself

1. What is formed when sulfuric acid reacts with sodium hydroxide?
2. What is the reaction between an acid and a base called?
3. Which ions are found in all acids?
4. What sort of salt does nitric acid produce?

⭐ Stretch Yourself

1. What is produced when acids react with metal oxides?

Metal Carbonate Reactions

Metal Carbonates

Metal carbonates react with acids to form a salt, water and carbon dioxide:

Metal Carbonate + Acid → Salt + Water + Carbon Dioxide

A gas (carbon dioxide) is made so bubbles will be seen. The name of the salt produced depends on the acid and the metal carbonate used:
- Hydrochloric acid makes chloride salts.
- Sulfuric acid makes sulfate salts.
- Nitric acid makes nitrate salts.

Making Salts from Metal Carbonates

Acids can be neutralised by metal carbonates to form salts. Most metal carbonates are insoluble, so they are bases, but they are not alkalis. When acids are neutralised by metal carbonates, a salt, water and carbon dioxide are produced. This means that rocks, such as limestone, that contain metal carbonate compounds are damaged by acid rain.

The general equation for the reaction is:

Metal Carbonate + Acid → Salt + Water + Carbon Dioxide

The reactions between acids and metal carbonates are **exothermic**.

Build Your Understanding

Examples of reactions between acids and metal carbonates:

Zinc Carbonate + Sulfuric Acid → Zinc Sulfate + Water + Carbon Dioxide

$$ZnCO_3 + H_2SO_4 \rightarrow ZnSO_4 + H_2O + CO_2$$

Copper Carbonate + Hydrochloric Acid → Copper Chloride + Water + Carbon Dioxide

$$CuCO_3 + 2HCl \rightarrow CuCl_2 + H_2O + CO_2$$

Make sure you can apply this idea to other examples.

The diagram opposite shows how copper chloride salt is made.

The steps involved in the production of copper chloride are as follows:
- Copper carbonate is added to hydrochloric acid until all the acid is used up.
- Any unreacted copper carbonate is filtered off.
- The solution of copper chloride and water is poured into an evaporating basin.
- The basin is heated gently until the first crystals of copper chloride start to appear.
- The solution is then left in a warm place for a few days to allow the remaining copper chloride to crystallise.

Naming Other Salts

Ethanoic acid is neutralised to form ethanoate salts. Phosphoric acid is neutralised to form phosphate salts.

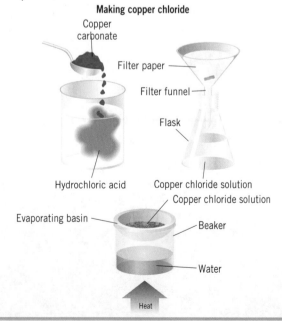

Making copper chloride

Identifying Carbonates

Metal **carbonates** react with dilute hydrochloric acid to form a salt, water and carbon dioxide gas.

To prove the gas produced is carbon dioxide, place a drop of limewater (calcium hydroxide $Ca(OH)_2$ solution) on a glass rod. If carbon dioxide is present the limewater turns cloudy:

Copper + Hydrochloric → Copper + Water + Carbon
Carbonate   Acid   Chloride   Dioxide

Symbol equation:

$CuCO_3(s) + 2HCl(aq) \rightarrow CuCl_2(aq) + H_2O(l) + CO_2(g)$

Ionic equation:

$CuCO_3(s) + 2H^+(aq) \rightarrow Cu^{2+}(aq) + H_2O(l) + CO_2(g)$

When copper carbonate is heated it decomposes to form copper oxide and carbon dioxide. This can be identified by a distinctive colour change: copper carbonate is green and copper oxide is black.

Testing for carbon dioxide

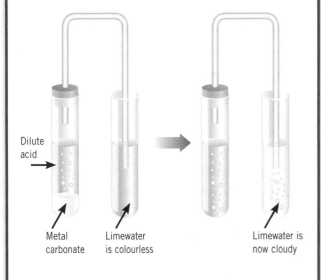

Dilute acid → Metal carbonate | Limewater is colourless | Limewater is now cloudy

Build Your Understanding

Dilute **sodium hydroxide** solution can be used to test for the presence of some transition metal ions in solution. The sodium hydroxide solution is added dropwise to the solution of the transition metal compound. The colour of the precipitate formed is used to identify the transition metal ion.

Transition Metal Ion	Results
Copper (II), Cu^{2+}	Blue precipitate of copper (II) hydroxide, $Cu(OH)_2$
Iron (II), Fe^{2+}	Green precipitate of iron (II) hydroxide, $Fe(OH)_2$ This turns brown as the Fe^{2+} ions are oxidised to Fe^{3+} ions

Sodium hydroxide can also be used to identify aluminium and calcium ions in solutions.

Metal Ion	Results
Aluminium, Al^{3+}	White precipitate that *does* dissolve in excess sodium hydroxide to form a colourless solution
Calcium, Ca^{2+}	White precipitate that *does not* dissolve in excess sodium hydroxide

To test for the presence of **ammonium ions** in a compound, add sodium hydroxide solution and then warm the resulting mixture. Then test for the presence of ammonia gas.

❓ Test Yourself

1. Which sort of salts does hydrochloric acid make?
2. Which sort of salts does sulfuric acid make?
3. Which sort of salts does nitric acid make?
4. What is the test for the gas carbon dioxide?

⭐ Stretch Yourself

1. A sample of copper carbonate was reacted with hydrochloric acid. Give the symbol equation for this reaction.

The Electrolysis of Sodium Chloride Solution

Sodium Chloride

Sodium chloride (**common salt**) is an important resource. It is an **ionic compound** formed from the **combination** of a group 1 metal (sodium) and a group 7 non-metal (chlorine). Sodium chloride is dissolved in large quantities in **seawater**.

Salt can be obtained by **mining** rocks or from allowing seawater to **evaporate**; the method used depends on how the salt is to be used and how pure it needs to be. **Quarrying** salt can have a dramatic impact on the environment. **Rock salt** (unpurified salt) is often used on icy roads. The salt lowers the freezing point of water from 0 °C to about −5 °C. Sprinkling rock salt on roads means that any water present will not freeze to form ice unless the temperature is very low.

Salt is also used in cooking, both to **flavour** food and as a **preservative**. However, eating too much salt can increase blood pressure and the chance of heart disease and strokes occurring. **Food packaging** may contain information on the sodium levels in a food.

Sodium ions may come from several sources, including sodium chloride salt. Government bodies, such as the Department for Health, produce guidelines to inform the public about the effects of different foods. A solution of sodium chloride in water is called brine.

To much salt can affect blood pressure

Electrolysis

The **electrolysis** of concentrated sodium chloride solution is an important industrial process and produces three useful products: chlorine, hydrogen and sodium hydroxide. The electrodes are made of inert materials so they do not react with the useful products made during the electrolysis reaction. The substance broken down is called the **electrolyte**.

Build Your Understanding

- During electrolysis, hydrogen ions, H^+, are attracted to the negative electrode where they pick up electrons to form hydrogen atoms, which pair up to make hydrogen, H_2, molecules:

  Hydrogen ions + Electrons → Hydrogen molecules

  $$2H^+ + 2e^- \rightarrow H_2$$

- Chloride ions, Cl^-, are attracted to the positive electrode where they release electrons to form chlorine atoms, which pair up to make chlorine molecules:

  Chloride ions − Electrons → Chlorine molecules

  $$2Cl^- - 2e^- \rightarrow Cl_2$$

- A solution of sodium hydroxide, $NaOH$, is also produced.

Each of these products can be used to make other useful materials. When there is a mixture of ions, such as in this case, the products that are formed depends on the reactivity of the elements involved. Electrolysis can also be used to electroplate objects. This can protect surfaces from corrosion and make them more attractive. Copper and silver plating are both produced by electrolysis.

Oxidation and Reduction

In the electrolysis of a concentrated sodium chloride solution:

- Hydrogen ions are **reduced** to hydrogen molecules; the hydrogen ions both gain an electron to form a hydrogen molecule.
- Chloride ions are **oxidised** to chlorine molecules; the two chloride ions both lose an electron to form a chlorine molecule.

Reduction and oxidation reactions must always occur together, so they are sometimes referred to as **redox** reactions.

Useful Products from the Electrolysis of Sodium Chloride

Chlorine is used:
- To make **bleach**.
- To sterilise water.
- To produce hydrochloric acid.
- In the production of PVC.

Hydrogen is used in the manufacture of margarine.

Sodium hydroxide is an alkali used in paper making and in the manufacture of many products including soaps and detergents, and rayon and acetate fibres.

Electrolysis of Molten Sodium Chloride

Solid sodium chloride does not conduct electricity because its ions cannot move. However, if sodium chloride is heated until it becomes **molten**, the sodium ions and chloride ions can move and electrolysis can occur.

Build Your Understanding

During the electrolysis of molten sodium chloride, the ions move towards the oppositely charged electrodes. Sodium, Na^+, ions (cations) are attracted to the negative electrode (cathode) where they pick up electrons to form sodium, Na, atoms:

Sodium ion + Electron → Sodium atom

$$Na^+ + e^- \rightarrow Na$$

Chloride, Cl^-, ions (anions) are attracted to the positive electrode (anode) where they release electrons to form chlorine atoms and then molecules:

Chloride ions − Electrons → Chlorine molecules

$$2Cl^- - 2e^- \rightarrow Cl_2$$

Make sure you can apply these ideas to other examples.

? Test Yourself

1. Which groups in the periodic table do sodium and chlorine belong to?
2. Where is sodium chloride found?
3. What is chlorine used for?
4. Why does solid sodium chloride not conduct electricity?
5. What is a solution of sodium chloride called?

★ Stretch Yourself

1. Give the symbol equations to show what happens at both the electrodes during the electrolysis of molten sodium chloride.
2. Name a health problem associated with eating too much salt.

Relative Formula Mass and Percentage Composition

Why Relative Atomic Mass is Used

Relative atomic mass (RAM or A_r) is used to compare the masses of different atoms. The relative atomic mass of an element is the average mass of its **isotopes** compared with an atom of carbon-12.

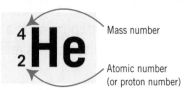

Relative Formula Mass

The **relative formula mass (RFM or M_r)** of a substance is worked out by adding together the relative atomic masses of all the atoms in the ratio indicated by the formula.

Example 1: For nitrogen, N_2:

$$N_2$$
$$(2 \times 14) = 28$$

The relative formula mass of N_2 is 28. Nitrogen molecules contain a triple covalent bond, which is very strong. This makes nitrogen molecules very stable.

Example 2: For carbon dioxide, CO_2:

$$CO_2$$
$$12 + (2 \times 16) = 44$$

The relative formula mass of CO_2 is 44.

Example 3: For water, H_2O:

$$H_2O$$
$$(2 \times 1) + 16 = 18$$

The relative formula mass of H_2O is 18.

The **molar mass** of a substance is its relative formula mass in grams. The units for molar mass are **g/mol**.

Build Your Understanding

The relative formula mass of a substance in grams is known as 1 **mole** of the substance. This is also called the molar mass. 1 mole of CO_2 is 44 g and 1 mole of H_2O is 18 g. The number of moles of a substance present can be calculated using this formula:

$$\text{Number of moles} = \frac{\text{mass of sample}}{\text{relative formula mass of the substance}}$$

Example 1: How many moles are there in 9 g of water?

The relative formula mass of water is 18 so the molar mass is 18 g.

$$\text{Number of moles} = \frac{9}{18} = 0.5$$

There are 0.5 moles in 9 g of water.

Build Your Understanding (cont.)

Example 2: What is the mass of 0.5 moles of nitrogen, N_2?

The relative formula mass of nitrogen is 28.

Mass of sample = number of moles × relative formula mass of the substance

= 0.5 × 28 = 14 g

The mass of 0.5 moles of nitrogen is 14 g.

Percentage Composition

Compounds consist of atoms of two or more different elements that have been chemically joined together. The percentage composition of an element in the compound can be calculated using this formula:

Percentage mass of an element in a compound = $\dfrac{\text{relative atomic mass} \times \text{no. of atoms}}{\text{relative formula mass}} \times 100\%$

Example 1: Ammonium nitrate is used as a fertiliser. Plants absorb fertiliser through their roots, so fertilisers must be soluble. Find the percentage composition of nitrogen in this compound:

- RAM of N = 14
- RAM of H = 1
- RAM of O = 16
- RAM of S = 32

The formula mass of ammonium nitrate, NH_4NO_3, is:

$$NH_4 \quad NO_3$$
$$14 + (4 \times 1) + 14 + (3 \times 16) = 80$$

Percentage of nitrogen = $\dfrac{14 \times 2}{80} \times 100\%$

= 35%

The percentage of nitrogen in ammonium nitrate is 35 per cent.

Example 2: Ammonium sulfate is also used as a fertiliser. Find the percentage of nitrogen in this compound:

$$(NH_4)_2 \quad SO_4$$
$$[14 + (4 \times 1)] \times 2 + 32 + (4 \times 16)$$

Percentage of nitrogen = $\dfrac{14 \times 2}{132} \times 100\%$

= 21.2%

The percentage of nitrogen in ammonium sulfate is 21.2 per cent.

Test Yourself

1. Why is relative formula mass used in science?
2. How is the relative formula mass of a substance calculated?
3. Find the relative formula mass of a nitrogen molecule, N_2.
4. Find the relative formula mass of carbon dioxide, CO_2.

Stretch Yourself

1. What is the relative formula mass of a substance in grams known as?
2. What is the mass of 1 mole of H_2O?

Calculating Masses

Calculating the Mass of Products

The masses of **products** and **reactants** can be worked out using the **balanced equation** for the reaction.

Example: What mass of water is produced when 8 g of hydrogen is burned?

Relative atomic masses:

H = 1, O = 16

First, write down what happens during the reaction as a word equation:

Hydrogen + Oxygen → Water

Then write it as a balanced symbol equation:

$2H_2 + O_2 → 2H_2O$

Next, calculate the **relative formula mass** of a hydrogen molecule and a water molecule.

The relative formula mass of hydrogen, H_2:

$$H_2$$
$$2 \times 1 = 2$$

The relative formula mass of water, H_2O:

$$H_2O$$
$$(2 \times 1) + 16 = 18$$

It is now possible to calculate the number of moles in 8 g of hydrogen.

$$\frac{8}{2} = 4 \text{ moles}$$

Next, examine the balanced symbol equation. Every 2 moles of hydrogen makes 2 moles of water. This means 4 moles of hydrogen will produce 4 moles of water.

Finally, work out the mass of 4 moles of water by rearranging the moles equation:

Mass of sample = number of moles × relative formula mass

= 4 × 18
= 72

This shows that if 8 g of hydrogen is burned completely, 72 g of water vapour will be produced.

Build Your Understanding

The equation for a reaction can also be used to calculate how much of the reactants should be used to produce a given amount of the product.

Example: What mass of magnesium should be used to produce 60 g of magnesium oxide? Relative atomic masses:

Mg = 24, O = 16

First, write down what happens during the reaction as a word equation:

Magnesium + Oxygen → Magnesium Oxide

Then write it as a balanced symbol equation:

$2Mg + O_2 → 2MgO$

Next, calculate the relative formula mass of magnesium oxide.

The relative formula mass of magnesium oxide, MgO:

$$MgO$$
$$24 + 16 = 40$$

It is now possible to calculate the number of moles in 60 g of magnesium oxide.

$$\frac{60}{40} = 1.5 \text{ moles}$$

Build Your Understanding (cont.)

Next, examine the balanced symbol equation. To make 2 moles of magnesium oxide, 2 moles of magnesium are needed. So, to make 1.5 moles of magnesium oxide, 1.5 moles of magnesium are needed.

Finally, work out the mass of 1.5 moles of magnesium oxide by rearranging the moles equation.

Mass of sample = number of moles × relative formula mass
= 1.5 × 24
= 36

This shows that to make 60 g of magnesium oxide, 36 g of magnesium should be burned.

Percentage Yield

The amount of product made in a reaction is called the **yield**. Although atoms are never gained or lost during a chemical reaction, the yield of a reaction can be less than predicted:

- The reaction is **reversible** and does not go to completion.
- Some of the product is lost during **filtering**, **evaporation**, when transferring liquids or during **heating**.
- There may be **side-reactions** occurring that produce other products.

The amount of product actually made compared with the maximum calculated yield is called the **percentage yield**. A 100 per cent yield means no product has been lost; a 0 per cent yield means no product has been made:

$$\text{Percentage yield} = \frac{\text{mass of product}}{\text{maximum calculated yield}} \times 100\%$$

Scientists try to choose reactions with a high percentage yield or **high atom economy**. This contributes towards **sustainable development** by reducing waste. A 100 per cent atom economy means that all the reactant atoms have been made into the desired products.

Waste products are undesirable as they cannot be sold for profit, their disposal can be costly and cause environmental and social problems.

Finding the Empirical Formula

The **empirical formula** of a compound is the simplest whole number ratio of the atoms it contains.

Example: Find the empirical formula of magnesium oxide formed when 12 g of magnesium reacts with 8 g of oxygen. Deal with the magnesium and oxygen separately.

The simplest ratio of magnesium atoms to oxygen atoms is 1 : 1 so the empirical formula is MgO.

	Mg	O
State the number of grams that combine	12	8
Change the grams to moles (divide by A_r)	$\frac{12}{24}$	$\frac{8}{16}$
This is the ratio in which the atoms combine	0.5	0.5
Get the ratio into its simplest form	1	1

❓ Test Yourself

1. What mass of water vapour is produced when 4 g of hydrogen is burned?
2. What mass of water vapour is produced when 16 g of hydrogen is burned?

⭐ Stretch Yourself

1. Consider the equation below:
$CaCO_3 \rightarrow CaO + CO_2$
If 5.0 g of calcium carbonate is heated fiercely, what mass of calcium oxide is produced?

Rates of Reaction

Slow and Fast Reactions

The **rate of reaction** is equal to either:
- The amount of reactant used up divided by the time taken.
- The amount of product made divided by the time taken.

Rusting is an example of a reaction that happens very slowly, while **combustion** reactions and explosions happen very quickly. **Explosions** produce a large volume of **gaseous products**. Factories that produce fine powders, such as custard powder, have to be very careful to prevent explosions from occurring.

Measuring Rates of Reaction

The method chosen to follow the rate for a particular reaction depends on the reactants and products involved. When sodium thiosulfate reacts with hydrochloric acid, one of the products is a precipitate of sulfur. The rate of reaction can be followed using a light sensor and a data logger to measure how quickly sulfur is being made.

A chemical reaction can only occur if the reacting particles collide with enough energy to react. This is called the **activation energy**. If the particles collide but do not have the minimum energy to react, the particles just bounce apart without reacting.

Measuring the rate of reaction

The rate of a chemical reaction can be measured by:
- How fast the products are being made.
- How fast the reactants are being used up.

The graph shows the amount of product made in two experiments. The lines are steepest at the start of the reaction in both experiments. The lines start to level out as the reactants get used up.

When the line becomes horizontal the reaction has finished. The graph shows that experiment A has a faster rate of reaction than experiment B. However, both experiments produce the same amount of product.

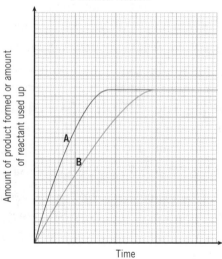

A graph to show how quickly a product is made in a chemical reaction.

When analysing graphs, the reaction is over when the graph levels out. Reactions stop when one of the reactants is all used up; this reactant is called the limiting reactant. The other reactants may not be completely used up and are said to be in excess. The amount of product made depends on the amount of reactant used up; the amount of product made is directly proportional to the amount of reactant used. The rate of reaction is measured using units of g/s or g/min or cm^3/s or cm^3/min.

Temperature

If the temperature is increased, the reactant particles move more quickly. Increasing the temperature increases the rate of reaction because:
- The particles collide more often.
- The particles have more energy when they collide.
- When they collide, the collision is more likely to lead to a reaction taking place between them.

Catalysts

A **catalyst** increases the rate of reaction, but is not itself used up during the reaction. Only a small amount of catalyst is needed to catalyse a large amount of reactants. Catalysts are specific to certain reactions. Reactions stop when one of the reactants is all used up. Catalysts offer an alternative reaction pathway with a lower activation energy.

Increasing the Surface Area

For a reaction to occur, the particles have to **collide**. The greater the surface area the more chance of the reactant particles colliding and the faster the rate of reaction.

With a small surface area (large pieces) the rate of reaction is slow. The particles collide less often. With a large surface area (small pieces) the rate of reaction is higher. The particles collide more often.

Large Particles	Small Particles
• Small surface area • Fewer collisions • Reaction rate is slow	• Large surface area • More collisions • Reaction rate is faster

✓ Maximise Your Marks

Remember, small pieces have a large surface area. The dust caused by fine powders, such as custard powder or flour, can burn explosively because of the large surface area of its particles.

Concentration and Pressure

For a reaction to take place, the reactant particles have to collide. If the concentration is increased there are more reactant particles in the solution. Increasing the concentration increases the rate of reaction because the particles collide more often.

For gases, increasing the pressure has the same effect as increasing the concentration of dissolved particles in solutions. At low pressure the rate of reaction slows down because the particles collide less often. At higher pressure the rate of reaction speeds up because the particles collide more often.

The concentrations of solutions are given in units of moles per cubic decimetre, $mol\ dm^{-3}$. Equal volumes of solutions of the same molar concentration contain the same number of moles of solute.

Increasing the pressure of gases increases the rate of reaction

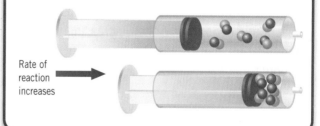

Rate of reaction increases

? Test Yourself

1. What happens to the rate of reaction if the concentration of reactants is increased?
2. What happens to the rate of reaction if the pressure of gaseous reactants is increased?
3. How does a catalyst affect the rate of reaction?

★ Stretch Yourself

1. Explain two ways in which increasing the temperature increases the rate of a chemical reaction.

Reversible Reactions

Simple Reversible Reactions

Not all reactions go to completion. Many chemical reactions are **reversible**; they can proceed both forwards and backwards.
If A and B are reactants and C and D are products, a reversible reaction can be summed up as:

A + B $\rightleftharpoons$ C + D

The two reactants, A and B, can react to make the products C and D; at the same time, C and D can react together to produce A and B.

Dynamic Equilibrium

If a reversible reaction takes place inside a closed system (where nothing can enter or leave), an **equilibrium** will eventually be reached.

It will be a **dynamic equilibrium**: both the forwards and the backwards reactions are taking place at exactly the same rate.

The conditions will affect the position of equilibrium, that is, how much reactant and product are present at equilibrium.

If the forwards reaction is exothermic then increasing the temperature will decrease the amount of product made. If the forwards reaction is endothermic then increasing the temperature will increase the amount of product made.

Build Your Understanding

In a dynamic equilibrium, both the forwards and the backwards reactions are still happening. As they happen at the same rate there is no overall change in the concentrations of the reactants or the products.

Build Your Understanding (cont.)

If the forwards reaction is exothermic (gives out energy), then the backwards reaction is endothermic (takes in energy). The amount of energy given out by the forwards reaction must be the same as the amount of energy taken in by the backwards reaction.

Example: Hydrated copper (II) sulfate reactions

First, the hydrated (with water) copper sulfate is heated to make anhydrous (without water) copper sulfate. Then water is added to the anhydrous copper sulfate to produce hydrated copper sulfate.

Copper (II) sulfate reactions

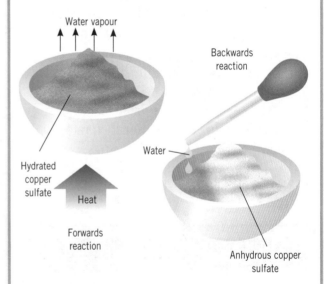

In the forwards reaction, the hydrated copper sulfate takes in energy as it is heated. This is an endothermic reaction:

**Hydrated Copper → Anhydrous Copper + Water
Sulfate (blue)   Sulfate (white)**

In the backwards reaction, energy is given out when water is added to the anhydrous copper sulfate. This is an exothermic reaction:

**Anhydrous Copper + Water → Hydrated Copper
Sulfate (white)   Sulfate (blue)**

Reactions Involving Gases

If a reaction involves gases then the pressure may affect the **yield** of the reaction.

First, count the number of gas molecules on the left-hand side and the right-hand side of the equation:

Reactants → Products

Fewer gas molecules → More gas molecules

Increasing the pressure decreases the yield of the product:

Reactants → Products

More gas molecules → Fewer gas molecules

Increasing the pressure increases the yield of the product.

Contact Process

The Contact process is used in the manufacture of sulfuric acid, H_2SO_4.

First, sulfur is burned in air to produce sulfur dioxide:

$S(s) + O_2(g) \rightarrow SO_2(g)$

Then the sulfur dioxide is reacted with more oxygen to form sulfur trioxide.

$2SO_2(g) + O_2(g) \rightleftharpoons 2SO_3(g)$

The forward reaction is exothermic.

Reaction Conditions

The raw materials are sulfur, water and oxygen. The reaction is carried out under the following conditions:
- Vanadium (V) oxide.
- Pressure of 1 atmosphere.
- Temperature of 450 °C.

Build Your Understanding

The vanadium (V) oxide catalyst is used because it increases the rate of reaction, which helps to reduce production costs.

A moderate temperature of 450 °C is chosen. This gives both a reasonable rate of reaction and a reasonable yield of the product.

The forwards reaction is exothermic so a higher temperature would give a faster rate of reaction but a lower yield of sulfur trioxide.

A lower temperature would give a higher yield of sulfur trioxide but a lower rate of reaction.

A higher pressure would increase the yield of the reaction because it would favour the forwards reaction, which decreases the number of gaseous molecules. However, it is expensive to maintain high pressures and, as the yield is already around 95 per cent, it is not necessary.

Finally, the sulfur trioxide is reacted with sulfuric acid to produce oleum, $H_2S_2O_7$:

$SO_3 + H_2SO_4 \rightarrow H_2S_2O_7$

The oleum is then reacted with water to form more sulfuric acid:

$H_2S_2O_7 + H_2O \rightarrow 2H_2SO_4$

? Test Yourself

1. What is special about a reversible reaction?
2. What is a closed system?
3. What is a dynamic equilibrium?

★ Stretch Yourself

1. The Contact process is used in the production of sulfuric acid.
    a) Name the catalyst used in the Contact process.
    b) Explain why a moderate temperature of 450 °C is used in this reaction.

Energy Changes

Energy Changes and Chemical Reactions

During chemical reactions atoms are rearranged as old **bonds** are broken and new bonds are made. **Energy** is required to break bonds and is released when new bonds are formed.

If, overall, energy is given to the surroundings, the reaction is described as **exothermic**. If, overall, energy is taken from the surroundings, the reaction is described as **endothermic**.

To work out whether a reaction is endothermic or exothermic, scientists measure the temperature of the chemicals before the reaction and then after the reaction. If the temperature has increased the reaction is exothermic; if the temperature has decreased the reaction is endothermic.

Exothermic and Endothermic Reactions

Burning methane is an example of an exothermic reaction. Rusting, explosions and neutralisation reactions are also exothermic. Self-heating cans and hand warmers make use of exothermic reactions.

The thermal decomposition of limestone is an example of an endothermic reaction. Photosynthesis and dissolving ammonium nitrate in water are also endothermic processes. Some sports injury packs make use of endothermic reactions.

Calorimetry – Burning Fuels

In **exothermic** reactions, more energy is given out when new bonds are formed than when the old bonds were broken. In **endothermic** reactions, more energy is taken in to break bonds than is released when new bonds are formed.

Scientists use **calorimetry** to compare the amount of energy released when fuels and foods are burned.

A measured amount of water is placed into a boiling tube and its temperature is measured using a thermometer. The fuel sample is then burned under the boiling tube containing the water, and the water is gradually warmed up. The temperature of the water is measured at the end of the experiment when the flame is extinguished.

The chemical energy that had been stored in the sample is released as thermal energy when the sample is burned. The greater the change in the temperature of the water the more energy was stored in the fuel.

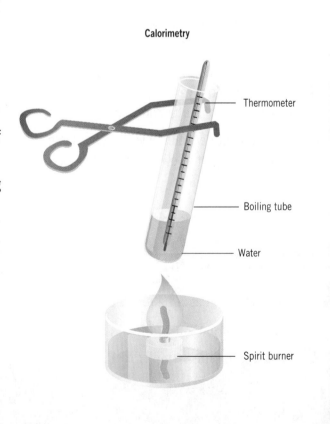

Calorimetry
- Thermometer
- Boiling tube
- Water
- Spirit burner

Build Your Understanding

Specific heat capacity (SHC) is the amount of energy needed to increase the temperature of 1 g of the substance by 1 °C.

Energy = mass × specific heat × change in change   capacity   temperature

The SHC of water is 4.18 J/g/°C.
1 cm^3 of water has a mass of 1 g.

Example: A 2 g sample of food made the temperature of 5 cm^3 of water increase by 6 °C.

Energy = mass × specific heat × change in
change   capacity   temperature

       = 5 g × 4.18 J/g/°C × 6 °C
       = 125.4 J

In these equations the mass is the mass of the water, not the mass of the fuel or the chemicals used.

Although energy is usually measured in joules, other units, such as kilojoules or calories, can be used:

- 1000 joules = 1 kilojoule
- 1 calorie = 4.2 joules

To be able to compare the energy content of different samples, energy values are normally given for the same amount of substance: joules per gram, kilojoules per gram or calories per gram.

The energy values for the example are given below using different units:

Joules per gram = $\frac{125.4 \text{ J}}{2 \text{ g}}$ = 62.7 J/g

Kilojoules per gram = $\frac{62.7 \text{ J}}{1000}$ = 0.0627 kJ

Calories per gram = $\frac{62.7 \text{ J}}{4.2 \text{ J}}$ = 14.9 calories

Boost Your Memory

Remember, combustion reactions are exothermic. Try making a list of all the other exothermic reactions you have learned about.

Calorimetry in Solutions

Calorimetry experiments can be used to work out the energy released by chemical reactions in solution. In the example below, a more reactive metal, iron, displaces a less reactive metal, copper, from a solution of copper sulfate.

Temperature at the start of the reaction	15 °C
Temperature at the end of the reaction	33 °C
Change in temperature	18 °C

The specific heat capacity of copper sulfate solution = 4.18 J/g/°C.

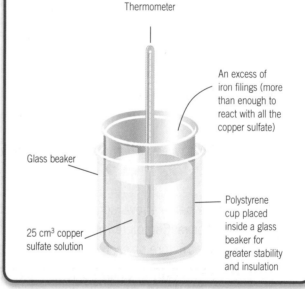

Measuring the temperature change of a displacement reaction

❓ Test Yourself

1. The temperature increases during a chemical reaction. What sort of reaction has taken place?

2. How do you know that burning coal is an exothermic reaction?

⭐ Stretch Yourself

1. A sample of sodium hydroxide pellets is dissolved in 25.0 g of water. The initial temperature of the water was 18 °C and the final temperature was 35 °C. The specific heat capacity of water is 4.18 J/g/°C. How much energy is given out in the process? Give your answer in kilojoules to three significant figures.

Practice Questions

Complete these exam-style questions to test your understanding. Check your answers on page 126-127. You may wish to answer these questions on a separate piece of paper.

1 The diagrams below show two atoms of the element oxygen. Atoms of oxygen contain three different types of particle. The particles in the atoms have been represented by the symbols **x**, ○ and ●.

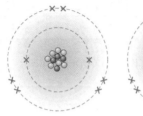

Oxygen 18

Oxygen 16

a) What type of particles do the crosses on the diagrams represent? (1)

b) i) What is the centre of an atom called? (1)

ii) Which types of particles do the dots in the centre of the oxygen atoms represent? (1)

c) The table below shows some information about the two oxygen atoms in the diagrams shown above. Complete the table to show the number of protons, neutrons and the electron structure of each of the oxygen atoms. (5)

	Number of Protons	Number of Neutrons	Electron Structure
$^{16}_{8}O$		8	
$^{18}_{8}O$			

2 Barium reacts with water to form barium hydroxide and hydrogen.

$Ba(s) + 2H_2O(l) \rightarrow Ba(OH)_2(aq) + H_2(g)$

0.69 g of barium was used in this reaction.

a) How many moles of barium were added to the water? Show your working. (2)

b) Calculate the number of moles of water that would be required to react with all the barium metal. (1)

3 a) Aluminium can be extracted from aluminium oxide. What is the name given to this process? (1)

b) Name the main ore of aluminium. (1)

c) Name the second ore of aluminium that is also used in the extraction of aluminium. (1)

d) Why is the second ore of aluminium used? (1)

e) During the electrolysis of aluminium oxide, the aluminium ions move. Which electrode do these ions move towards? (1)

f) Which of these words best describes what happens to aluminium ions during electrolysis? Tick one box. (1)

☐ Displacement   ☐ Oxidation
☐ Reduction   ☐ Neutralisation

4 The electrolysis of concentrated sodium chloride solution is an important industrial process.

a) Name the gas produced at the positive electrode in this process. (1)

b) Give one use of this gas. (1)

c) Name the gas produced at the negative electrode in this process. (1)

d) Give one use of this gas. (1)

e) Name the other chemical made in this reaction. (1)

f) Give one use of this chemical. (1)

How well did you do?

| 0–7 Try again | 8–14 Getting there | 15–19 Good work | 20-23 Excellent! |

Answers

Biology

Cells and Organisation Pages 4-5
Test Yourself
1. Cell wall, vacuole, chloroplasts.
2. Cell sap can have many functions. It helps support the plant.
3. Organs.

Stretch Yourself
1. 10 mm wide so 200 times (10 mm / 0.005 mm = 200 mm)
2. The cells become so specialised that they cannot take on other roles and take the function of those lost.

DNA and Protein Synthesis Pages 6-7
Test Yourself
1. Because two helices are twisted into a spiral shape.
2. The structure has a different order of amino acids.
3. Hydrogen bonds between the bases, C with G and A with T.
4. Each sequence of three bases codes for one amino acid.

Stretch Yourself
1. 16%
2. The double helix unwinds and the base pairs separate. This allows mRNA bases to pair with the DNA bases producing a complimentary copy to pass to the ribosomes.

Proteins and Enzymes Pages 8-9
Test Yourself
1. Proteins are needed to make key structures inside the body, e.g. bone and so without protein growth will be limited.
2. To allow reactions to be fast enough at body temperature.
3. The lock is the enzymes active site and the key is the substrate.
4. The temperature at which the reaction occurs at the fastest rate.

Stretch Yourself
1. Lipase has a particular shaped active site that fats will fit into, but not proteins.
2. Vinegar is acidic and so the pH would be too low so the enzymes of the decay organisms would not work.

Cell Division Pages 10-11
Test Yourself
1. One
2. In the ovaries and testes.
3. 46
4. Meiosis introduces more variation; meiosis makes four cells but mitosis makes two; meiosis makes cells with half the number of chromosomes, but mitosis produces cells that have the same number as the parent cells. Mitosis produces cells that are genetically identical to each other and to the parents – meiosis produces genetically different cells.

Stretch Yourself
1. It usually produces a new protein that does not work as well.
2. UV light can be absorbed by DNA and cause mutations.

Growth and Development Pages 12-13
Test Yourself
1. Answers might include muscle, blood or nerve cells.
2. A cell that has not yet differentiated and so can divide to produce any type of cell.
3. Growth becomes negative, i.e. more cells are dying than are being produced.
4. The meristems at the tips of the roots and shoots.

Stretch Yourself
1. The brain needs to develop first to control the other parts of the body.
2. This is destructive and it is difficult to tell when all the water has been removed; Organism has to be killed and all water removed in order to measure dry mass. Therefore it is hard to get an idea of growth over time.

Transport in Cells Pages 14-15
Test Yourself
1. The molecules from the 'stink bomb' move from a high concentration to a low concentration over the other side of the room. It diffuses.
2. Diffusion is faster in warm conditions because the molecules are moving faster.
3. A membrane that lets some molecules through but not others.
4. Water

Stretch Yourself
1. Accept answers between 0.16-0.18 moles
2. The air spaces between the soil particles have been filled with water, which takes the place of oxygen.

Respiration Pages 16-17
Test Yourself
1. We need to respire more to generate more heat to keep a constant body temperature, so more food is needed to provide glucose for respiration to produce this heat.
2. Respiration rate increases to supply extra energy for muscle contraction. More oxygen is therefore needed and more carbon dioxide needs to be removed.
3. Carbon dioxide.
4. Lactic acid is produced due to anaerobic respiration.

Stretch Yourself
1. 6/6 = 1.0
2. So that the temperature is at an optimum level for the yeast enzymes to produce alcohol.

Sampling Organisms Pages 18-19
Test Yourself
1. A population.
2. a) quadrat b) net c) pitfall trap.
3. 20 in 1m^2 so 2000 in the field.
4. One area of the field may not be representative of all the areas. It saves time.

Stretch Yourself
1. 210
2. Some animals can survive more time out of the water than others and so can live further up the shore.

Photosynthesis Pages 20-21
Test Yourself
1. To absorb the energy of sunlight.
2. Palisade mesophyll.
3. Stomata are structures containing leaf pores that allow gases in and out of the leaf.
4. Leaves have veins to supply water for photosynthesis and to take away the sugars that are produced from photosynthesis.

Stretch Yourself
1. It is released from water.
2. High temperatures will denature the enzymes that control the reactions of photosynthesis.

Food Production Pages 22-23
Test Yourself
1. Plants need nitrates to produce amino acids and therefore proteins.
2. Chlorophyll cannot be made, as chlorophyll contains magnesium.
3. Growing plants without the use of soil just in water containing nutrients.
4. The temperature is not low enough to completely stop the growth of microbes that cause decay.

Stretch Yourself
1. Often they kill the natural predators of the pests too and so increase the problem. They also contaminate the environment and reduce biodiversity.
2. It prevents loss of energy in pig movement therefore leaving more energy for growth.

Transport in Animals Pages 24-25
Test Yourself
1. Platelets clot the blood.
2. To fit more haemoglobin in, so they can carry more oxygen.
3. To stop the blood flowing backwards as the pressure is low.
4. Pulmonary artery.

Stretch Yourself
1. It contains deoxygenated blood (which is dark red, not blue).
2. When the left ventricle contracts some of the blood goes back into the left atrium rather than out into the aorta. This causes a backlog of blood in the veins coming back from the lungs.

Transport in Plants Pages 26-27
Test Yourself
1. In a plant stem water and minerals move upwards, towards the leaves.
2. The movement of dissolved food through the phloem.
3. Running down the centre.
4. To increase the surface area for water absorption.

Stretch Yourself
1. Because carbon dioxide must be allowed in for photosynthesis.
2. The guard cells lose water by osmosis and so the cells become flaccid, straightening up and closing the pore.

Digestion and Absorption Pages 28-29
Test Yourself
1. The starch is being digested into the sugar maltose by amylase.
2. The gall bladder stores bile.
3. Fatty acids and glycerol.
4. It is sweeter so less is needed.

Stretch Yourself
1. So that the acid does not damage it and the protease does not digest it.
2. Slows down the rate of absorption as there is a smaller surface area.

Practice Questions
1. All organisms release energy from food. This largely happens in the **mitochondria**. Cells take up water by osmosis because the **cell membrane** is partially permeable. The **vacuole** stores some sugars and salts. Plant cells are limited to how much water they can take up because the **cell wall** resists the uptake of too much water.
2. a) Pondweed lives underwater, makes bubbles of oxygen that can be counted, it is easy to obtain and photosynthesises readily.
   b) Oxygen.
   c) He uses the same piece of pondweed throughout.
   d) Use a more accurate timer, such as a stop watch.
3. a) On the chromosomes; in the nucleus.
   b) Each gene has a different order of DNA. The bases code for the order of amino acids in the protein.
   c) 1094/20 000 × 100 = 5.47%
   d) The liver – it uses the most genes and they code for different enzymes.
4. Starch molecules are too large to be able to pass into the bloodstream and so need to be **digested** first. This digestion begins in the **mouth**. An enzyme called **amylase** breaks down starch into maltose. Maltose is then digested into **glucose** in the small intestine. Absorption then occurs and this is speeded up by the presence of tiny projections on the wall of the small intestine called **villi**.
5. A = artery; B = vein; C = capillary

Physics

Distance, Speed and Velocity Pages 32-33
Test Yourself
1. Instantaneous speed is the speed at any instant. Average speed is calculated from the total distance travelled divided by time taken and the speed may have changed several times during the journey time.
2. $(100\text{ m} - 0\text{ m}) \div (30\text{ s} - 0\text{ s}) = 3.33\text{ m/s}$
3. Speed = 288 km ÷ 3 h = 96 km/h
4. $t = d \div s$; $t = 1.44$ km ÷ 12 m/s = 1440 m ÷ 12 m/s = 120 s (= 2 minutes)

Stretch Yourself
1. a) 0.9 km
   b) Between 1.5 and 2 hours (or on the way home from the shop)
   c) Because the graph is steepest at this point.

Speed, Velocity and Acceleration Pages 34-35
Test Yourself
1. 24 m/s ÷ 8 s = 3 m/s^2
2. The object is travelling at a steady/constant speed.
3. a) 0 m/s^2
   b) (0–20) m/s ÷ (80–60) s = –1 m/s^2

Stretch Yourself
1. Line goes below x axis similar to last part of the graph for the rolling ball, then horizontal, then straight line sloping upwards to the x axis again.
2. a) 20 m/s × (60 s – 40 s) = 400 m
   b) ½ × 20 m/s × (80 s – 60 s) = 200 m

Forces Pages 36-37
Test Yourself
1. 2 kg × 10 N/kg = 20 N
2. 3000 N – 900 N = 2100 N
3. 0 N because there is no change in motion.

Stretch Yourself
1. a) 0.1 kg × 10 kg/N = 1 N; b) 1/6 × 1 N = 0.17 N
2. 2500 N – 800 N – 1700 N = 0 N

Acceleration and Momentum Pages 38-39
Test Yourself
1. 1200 kg × 3 m/s^2 = 3600 N
2. 2 kg × 5 m/s = 10 kgm/s (or 10 Ns)
3. 0.2 kg × (–8 m/s) = –1.6 kgm/s (or –1.6 Ns)
4. 50 N × 12 s = 600 Ns

Stretch Yourself
1. a) Force = (1000 kg × 12 m/s) ÷ 0.002 s = 6 000 000 N
   b) Stopping force = (1000 kg × 12 m/s) ÷ 0.5 s = 24 000 N
   c) There would be less damage to the truck.

Pairs of Forces: Action and Reaction Pages 40-41
Test Yourself
1. The wall pushes on the girl with a force of 5 N.
2. Because there is little friction to push back on your foot and send you forward.
3. It pushes exhaust gases out the back, so there is an equal and opposite force pushing the rocket forward.
4. The balloon pushes air out of the back and an equal and opposite force on the balloon pushes it forward.

Stretch Yourself
1. 1st pair: The weight of the book and the equal and opposite gravitational attraction of the book on the Earth. 2nd pair: The contact force of the book pushing down on the table and the equal and opposite force of the table pushing up on the book.
2. They both stop (momentum = 0), so before the collision the momentum of the car is equal and opposite to the momentum of the truck.
   Car: mv = 0.5 kg × 4 m/s = 2 kgm/s
   Truck: MV = 2 kg × V V = –2 kgm/s ÷ 2 kg = –1 m/s (speed = 1 m/s)

Work and Energy Pages 42-43
Test Yourself
1. 1000 N × 200 m = 200 000 J
2. 12000 N × 30 m = 360 000 J
3. ½ × 900 kg × (15 m/s)^2 = 101 250 J
4. If there is friction some of the energy is not transferred from GPE to KE.

Stretch Yourself
1. a) 8000 N × 50 m = 400 000 J; b) 400 000 J

Energy and Power Pages 44-45
Test Yourself
1. When the frictional forces are so small they can be ignored, or in a vacuum.
2. The GPE is transferred to heating the air and the object.
3. Any of listed examples, e.g. driver has been drinking or poor condition of tyres.

Stretch Yourself
1. The cyclist is doing work against friction forces the energy is transferred as heat to the bicycle, road and air.
2. 3 × 30 mph so thinking distance = 3 × 9 m braking distance = 9 × 14 m stopping distance = 27 + 126 = 153 m

Electrostatic Effects Pages 46-47
Test Yourself
1. They have opposite charges, which attract.
2. Electrons
3. Two – positive and negative.
4. Nylon is a better insulator. Some of the charges are conducted away through the wool carpet.

Stretch Yourself
1. a) They are opposite
   b) Repel because they are the same, and they would get the same charge.
2. So that there is not a build up of charge which could cause a spark. A spark could ignite the fuel.

Uses of Electrostatics Pages 48-49
Test Yourself
1. The grids will attract (or repel) electrons from the smoke particles, which will be left with the same charge as the grid, so the plates need to have the opposite charge to attract the charged smoke particles.
2. They would be repelled and would not stick to the car body.
3. Better coverage or less wastage.
4. The shock could stop their heart.

Stretch Yourself
1. So that they will charge the smoke particles that come close to them, by attracting or repelling electrons.
2. Electrons flow from earth to neutralise the positive charge of the paint drops landing on the metal.

Electric Circuits Pages 50-51
Test Yourself
1. a) 0.8 A; b) 0.8 A
2. a) 400 mA; b) 900 mA

Stretch Yourself
1. 20 C
2. reading A = reading B + reading C

Voltage or Potential Difference Pages 52-53
Test Yourself
1. a) 2 V
   b) No, current is the same everywhere in a series circuit.
2. a) 9 V
   b) Only if the lamps are identical. They might draw a different current and have different brightness.

Stretch Yourself
1. 9 V × 10 C = 90 J
2. a) 5 × 1.5 V = 7.5 V; b) 1.5 V
3. They will last longer – there is five times as much stored charge.

Resistance and Resistors Pages 54-55
Test Yourself
1. 90 Ω
2. 0.06 A
3. a) 600 Ω; b) R_3 = 300 Ω

Stretch Yourself
1. The free electrons collide with the stationary positive ions giving them energy so they vibrate more, which means they are hotter.
2. Yes, because V/I = 150 Ω in each case, so R is constant when V and I change, V is proportional to I.

Special Resistors Pages 56-57
Test Yourself
1. Its temperature increases and so resistance increases.
2. It has too high a resistance and would get hot when current flowed.
3. Its resistance decreases when temperature increases.
4. A diode – it only allows current to pass in one direction.

Stretch Yourself
1. a) $V_1 = 5\text{ V} \times \dfrac{300\text{ Ω}}{(300\text{ Ω} + 200\text{ Ω})} = 3\text{ V}$
   b) $V_2 = 5\text{ V} \times \dfrac{200\text{ Ω}}{(300\text{ Ω} + 200\text{ Ω})} = 2\text{ V}$ or 5 V – 3 V = 2 V
2. a) decreases; b) decreases; c) increases.

The Mains Supply Pages 58-59
Test Yourself
1. It won't melt if the current gets too high – other wires may melt first causing a fire.
2. If the appliance becomes live it won't melt the fuse, so someone touching it could get a fatal shock.
3. a) 2500 W ÷ 230 V = 10.9 A
   b) 9 W ÷ 230 V = 0.04 A
   c) 300 W ÷ 230 V = 1.3 A
4. a) 13 A; b) 3 A; c) 3 A

Stretch Yourself
1. If they get wet/wires get cut/other fault the power supply will be cut off. When the fault is fixed the power can be switched back on without replacing the fuse.
2. a) (0.5 A)^2 × 100 Ω = 25 W
   b) (1 A)^2 × 100 Ω = 100 W

Atomic Structure Pages 60-61
Test Yourself
1. a) Protons, neutrons, electrons
   b) Protons, neutrons.
2. There are 8 protons in the nucleus (and 8 orbital electrons when the atom is neutral).
3. nitrogen-14 and nitrogen-16.

Stretch Yourself
1. $^{235}_{92}\text{U} \rightarrow {}^{231}_{90}\text{Th} + {}^{4}_{2}\text{He}$
2. $^{16}_{7}\text{N} \rightarrow {}^{16}_{8}\text{O} + {}^{0}_{-1}\text{e}$

Radioactive Decay Pages 62-63
Test Yourself
1. a) Alpha particle
   b) Alpha particles and beta particles.
2. a) ½; b) 1/16

Stretch Yourself
1. A helium atom.
2. a) 16000 ÷ 2 = 8000 Bq
   b) 16000 ÷ 2^3 = 2000 Bq
   c) 16000 ÷ 2^{10} = 16
   d) 16000 Bq × ½ × ½ × ½ × ½ × ½ = 500 Bq = 5 half-lives = 5 x 3 = 15h

Living with Radioactivity Pages 64-65
Test Yourself
1. 50% (or half)
2. Kill; damage; damage DNA; turn cancerous.
3. Her risk of getting cancer is increased, but we can't tell whether she will get cancer.

Stretch Yourself
1. Cornwall and Cumbria or other red areas on the map.

Uses of Radioactive Materials Pages 66-67
Test Yourself
1. a) e.g. smoke detector
   b) e.g. medical tracer
   c) e.g. cancer treatment.
2. Each of the beams is a low dose so it doesn't kill the cells, but when combined in the tumour the dose is high enough to kill the cells.

Stretch Yourself
1. Alpha is stopped by the smoke, beta and gamma are not.
2. There is insufficient radioactive carbon left in the egg, so it could be much older – you can't tell.

Nuclear Fission and Fusion Pages 68-69
Test Yourself
1. Fission is splitting large nuclei into two roughly equal parts, fusion is joining two small nuclei.
2. The plutonium-239 nucleus absorbs a neutron splits into two parts and a few extra neutrons. These neutrons are absorbed by more plutonium-239 nuclei which split – and so on.
3. To absorb neutrons and stop them causing more nuclei to fission. The rods are lowered or raised to change the number of neutrons absorbed and control the rate of reaction.
4. Stored under water in cooling tanks (high level waste).

Stretch Yourself
1. 0.1 g = 0.0001 kg × (3 × 10^8 m/s)^2 = 9 × 10^{12} J
2. The strong force.

Stars Pages 70-71
Test Yourself
1. The outer layers of a red giant star that has used up its helium and are drifting away into space in all directions.
2. A brown or black dwarf.
3. A supernova.
4. A red supergiant is more massive, and after helium fusion is complete it goes on to fuse other nuclei until it produces iron. Finally, it explodes as a supernova. A red giant stops after helium fusion is finished producing a planetary nebula and a white dwarf.

Stretch Yourself
1. By nuclear fusion, inside red giant and red supergiant stars.
2. Because no light can escape it, as there is such a strong gravitational attraction.

Practice Questions
1. a) It accelerates at a steady rate from 0 m/s to 14.0 m/s in 10 seconds. It travels at a steady speed for 25 seconds. It slows down at a steady rate to 8.5 m/s over 15 seconds
   b) i) (14 − 0) m/s ÷ 10 s = 1.4 m/s^2
   ii) 14 m/s
   iii) (8.5 m/s − 12 m/s) ÷ 10 s = −3.7 m/s ÷ 10 s = −0.37 m/s^2 (or (8.3 m/s−14 m/s) ÷ 15 s = −0.38 m/s^2] (answers between 0.37 m/s^2 and 0.4 m/s^2 are acceptable)
   c) The distance travelled during the first 10 seconds
   d) i) ½ (14 m/s × 10 s) = 70 m
   ii) 72.5 m + (14 m/s × 10 s) = 75 m + 290 m = 365 m
2. a) 20 mA
   b) 20 mA × 60 Ω = 1.2 V
   c) 3 V − 1.2 V = 1.8 V
   d) 50 mA − 20 mA = 30 mA
   e) 3 V ÷ 30 mA = 100 Ω
3. a) 8000 W ÷ 230 V = 34.8 A
   b) A very high current so this would heat cables, thick cables have lower resistance so less heating.
4. a) Sodium-24 has one extra neutron
   b) 15 hours
   c) 20 counts per minute
   d) i) No, because it would decay too fast and there would be no radioactivity left after a few weeks/ because it is a gamma emitter and people who came close to it would be irradiated
   ii) Yes, because it is a gamma emitter so the radiation from the pipe would reach the surface/ because it has a short enough half-life not to contaminate the site for a long time
5. You will be given an injection containing radioactive iodine-131. The iodine will be absorbed by your thyroid gland. The radioactive iodine nuclei will decay by emitting beta particles. Beta particles are ionising and they can kill living cells, so many of the beta particles will kill the tumour cells. In 8 days half of the iodine-131 will have gone and in 16 days there will be only a quarter left. After 80 days, which is 10 half-lives, there will be less than one thousandth left. To decide whether to have the treatment you must weigh up the benefits and the risks. The risk is that beta particles will hit healthy cells and kill or damage them. You can replace a few killed cells, but a damaged cell could mutate and cause a new cancer. This doesn't mean it will cause cancer, only that there is a risk. The benefit is that the beta particles will kill the tumour cells so that you are cured.

Chemistry

Atomic Structure Pages 74-75
Test Yourself
1. Protons and neutrons.
2. Electrons.
3. Charge +1, mass 1 amu.
4. Charge −1, mass negligible.

Stretch Yourself
1. As both elements are in group 2 of the periodic table they have a similar electron configuration; they both have two electrons in their outer shell.

Atoms and the Periodic Table Pages 76-77
Test Yourself
1. To check them against new evidence that is found.
2. They are in order of increasing atomic number.
3. a) Periods; b) Groups.

Stretch Yourself
1. Dalton predicted that atoms cannot be divided into simpler substances. It is now known that atoms are made of protons, neutrons and electrons. All atoms of the same element are the same. Scientists now know about isotopes. These are different forms of the same element that have the same number of protons and a different number of neutrons.

The Periodic Table Pages 78-79
Test Yourself
1. The periodic table.
2. It had not been discovered.

Stretch Yourself
1. 2+.

Chemical Reactions and Atoms Pages 80-81
Test Yourself
1. Atoms can be joined by sharing electrons or by giving and taking electrons.
2. Sodium and chromium.
3. It consists of two hydrogen atoms and one oxygen atom.
4. It consists of sodium atoms, nitrogen atoms and oxygen atoms in the ratio 1 : 1 : 3.

Stretch Yourself
1. a) KCl; b) NaBr.

Balancing Equations Pages 82-83
Test Yourself
1. Atoms cannot be created or destroyed during chemical reactions.
2. 2Na + Cl$_2$ → 2NaCl.
3. H$_2$ + Cl$_2$ → 2HCl.
4. It is a liquid.

Stretch Yourself
1. The reacting ions collide together and react very quickly.

Ionic and Covalent Bonding Pages 84-85
Test Yourself
1. Ions are atoms or groups of atoms with a charge.
2. Electrons are transferred.
3. Shared pairs of electrons.
4. 1+.

Stretch Yourself
1. a) Covalent bonding, with a double covalent bond between the two oxygen atoms.
   b) Ionic bonding between the positively charged sodium ions and the negatively charged chloride ions.

Ionic and Covalent Structures Pages 86-87
Test Yourself
1. a) Ionic compound.
   b) Giant covalent structure.
   c) Giant covalent structure.

Stretch Yourself
1. Sodium chloride contains sodium ions and chloride ions. The ions cannot move when it is solid, so solid sodium chloride does not conduct electricity. When the sodium chloride is dissolved in water the ions can move, so aqueous sodium chloride does conduct electricity.

Group 7 Pages 88-89
Test Yourself
1. Halogens.
2. Halide ions are formed when a halogen atom gains an electron.
3. Ionic compound.

Stretch Yourself
1. Chlorine + Potassium Iodide → Iodine + Potassium Chloride
   Cl$_2$ + 2KI → I$_2$ + 2KCl

New Chemicals and Materials Pages 90-91
Test Yourself
1. 1 nm–100 nm.
2. C$_{60}$.
3. In sea spray.

Stretch Yourself
1. The chemicals are made all the time. Raw materials are continuously added and new products are removed.

Plastics and Perfumes Pages 92-93
Test Yourself
1. Addition polymerisation.
2. Monomers.
3. They are cheaper than natural ingredients.

Stretch Yourself
1. The diagram has two carbon atoms joined by a double bond. Each carbon atom is also joined by single bonds to two fluorine atoms.
2. Unsaturated – there is a double bond between the two carbon atoms.
3. Perfumes must evaporate easily so they can travel through the air and be smelt.

Analysis Pages 94-95
Test Yourself
1. Modern methods are faster, more sensitive, more accurate, and smaller samples are needed.
2. Chromatography is used to separate out the components of a mixture.

Stretch Yourself
1. Green = $\frac{6.0 \text{ cm}}{8.0 \text{ cm}}$ = 0.75
   Blue = $\frac{5.0 \text{ cm}}{8.0 \text{ cm}}$ = 0.625

Metals Pages 96-97
Test Yourself
1. Metallic bonding is the attraction between the positive metal ions and the sea of negative, delocalised electrons.
2. They have delocalised electrons that can move.
3. Nickel and titanium.
4. A lot of energy is required to overcome the metallic bonds.

Stretch Yourself
1. Nitinol is used in some dental braces.
2. Superconductors only work below their critical temperatures; at present these temperatures are too low to be readily attainable.

Group 1 Pages 98-99
Test Yourself
1. Lithium, sodium and potassium.
2. One.
3. They have the same outer electron structure.
4. Ionic compounds.

Stretch Yourself
1. a) Potassium + Chlorine → Potassium Chloride
   2K + Cl$_2$ → 2KCl
   b) K → K$^+$ + e$^-$. Potassium has lost an electron so it is oxidised in this reaction.

Aluminium and Transition Metals Pages 100-101
Test Yourself
1. Electrolysis.
2. The middle section, between groups 2 and 3
3. The negative electrode

Stretch Yourself
1. Bauxite has a very high melting point and the addition of cryolite reduces this temperature and, therefore, the energy needed.
2. Many transition metals can form ions with different charges.

Chemical Tests Pages 102-103
Test Yourself
1. MgCl$_2$
2. Yellow.

Stretch Yourself
1. First, clean a flame test wire by placing it into the hottest part of a Bunsen flame. Next, dip the end of the wire into water and then into the salt sample. Finally, hold the salt in the hottest part of the flame and observe the colour seen.

Acids and Bases Pages 104-105
Test Yourself
1. H$^+$ ions.
2. OH$^-$ ions.
3. Hydrochloric acid, sulfuric acid and nitric acid are all strong acids. They are completely ionised in water.

Stretch Yourself
1. **a)** Both reactions would produce carbon dioxide/bubbles would be seen/the same volume of gas would be made in both experiments.
   b) The reaction involving hydrochloric acid would be faster because it is a strong acid while ethanoic acid is a weak acid.

Making Salts Pages 106-107
Test Yourself
1. Sodium Sulfate + Water.
2. Neutralisation.
3. Hydrogen, H^+, ions.
4. Nitrates.

Stretch Yourself
1. A salt and water.

Metal Carbonate Reactions Pages 108-109
Test Yourself
1. Chlorides.
2. Sulfates.
3. Nitrates.
4. Carbon dioxide turns limewater cloudy.

Stretch Yourself
1. $CuCO_3 + 2HCl \rightarrow CuCl_2 + H_2O + CO_2$

The Electrolysis of Sodium Chloride Solution Pages 110-111
Test Yourself
1. Groups 1 and 7.
2. It is found in seawater and in underground deposits.
3. It is used in bleach, to sterilise water, in the production of hydrochloric acid and PVC.
4. The ions cannot move.
5. Brine.

Stretch Yourself
1. $Na^+ + e^- \rightarrow Na$
   $2Cl^- - 2e^- \rightarrow Cl_2$
2. One of: high blood pressure; heart disease; stroke.

Relative Formula Mass and Percentage Composition Pages 112-113
Test Yourself
1. It is used to compare the masses of different compounds.
2. By adding together the relative atomic masses of all the atoms in the ratio indicated by the formula.
3. 28.
4. 44.

Stretch Yourself
1. 1 mole.
2. 18 g.

Calculating Masses Pages 114-115
Test Yourself
1. 36 g.
2. 144 g.

Stretch Yourself
1. 2.8 g.

Rates of Reaction Pages 116-117
Test Yourself
1. It increases.
2. It increases.
3. It increases.

Stretch Yourself
1. If the temperature is increased the particles move more quickly. This means the particles collide more often and, when they do collide, the collisions have more energy. As more collisions have a level of energy greater than the activation energy, the particles react more quickly.

Reversible Reactions Pages 118-119
Test Yourself
1. It can proceed in either direction.
2. It is where nothing can enter or leave.
3. It is where the rate of forwards and backwards reactions are the same, so there is no change in the overall concentrations of reactants or products.

Stretch Yourself
1. **a)** Vanadium (V) oxide.
   b) A higher temperature would give a faster rate of reaction but a lower yield of sulfur trioxide. A lower temperature would give a higher yield of sulfur trioxide but a lower rate of reaction. A moderate temperature gives both a reasonable rate of reaction and a reasonable yield of the product.

Energy Changes Pages 120-121
Test Yourself
1. Exothermic.
2. It releases lots of energy.

Stretch Yourself
1. Energy change = mass × specific heat capacity × change in temperature
   = 25 g × 4.18 J/g/°C × 17 °C
   = 1776.5 J
   = 1.7765 kJ
   = 1.78 kJ

Practice Questions
1. **a)** Electrons.
   b) i) The nucleus.
   ii) Protons and neutrons.

Number of Protons	Number of Neutrons	Electron Structure
8	8	2, 6
8	10	2, 6

2. **a)** $\dfrac{0.69}{137} = 0.005$ moles.
   b) 0.01 moles.
3. **a)** Electrolysis.
   b) Bauxite.
   c) Cryolite.
   d) It has a lower melting point and bauxite dissolves in molten cryolite.
   e) The negative electrode.
   f) Reduction.
4. **a)** Chlorine.
   b) One of: bleach; to sterilise water; PVC; hydrochloric acid.
   c) Hydrogen.
   d) It is used in the manufacture of margarine.
   e) Sodium hydroxide.
   f) One of: soaps; detergents; rayon; acetate.

Index

A
Absorption 28-29
Acceleration 34-35, 38-39
Acids 104-105
Action 40-41
Active Transport 14-15
Alkali metals 98-99
Aluminium 100-101
Amino acids 7
Analysis (chemical) 94-95
Atoms 60-61, 74-77, 80-81

B
Background radiation 64-65
Bacteria 4
Balanced forces 36-37
Bases 104-1-5
Blood 24
Blood vessels 24
Buckminster fullerene 90

C
Calculating mass 114-115
Calorimetry 120-121
Carbon dating 67
Catalysts 117
Cells 4, 10, 14-15
Chemical tests 102-103
Chromatography 94-95
Chromosomes 12-13
Circuit symbols 50-51
Circulation 25
Collisions 38-39
Concentration 117
Conductors 46-47, 51, 96
Conservation of energy 44-45
Conservation of mass 82
Contact process 119
Contamination 65
Copper 101
Covalent bonding 81, 84-87
Covalent structures 86-87

D
Development 12-13
Diamond 87
Differentiation 12
Diffusion 14-15
Digestion 28-29
Distance 32-33
Distance-time graphs 32-33
DNA 6-7, 10
Dynamic equilibrium 118

E
Earth connection 47
Ecosystems 19
Electric charge 46-47
Electric circuits 50-51
Electric current 51
Electrical power 59
Electrolysis 100-101, 110-111
Electrons 60
Electrostatic 46-49
Elements 74-75
Empirical formula 115
Endothermic reactions 118-120
Energy 16-17, 42-45, 120-121
Energy change 120-121
Enzymes 8-9, 28-29
Equations 81-83
Esters 93
Exothermic reactions 118-120

F
Fission 68-69
Flame tests 102
Food production 22-23
Force 38-39
Forces 36-39, 40-41
Formulae 80, 103
Friction 40-41
Fuses 58-59
Fusion 68-71

G
Graphite 87
Gravitational potential energy 42-43
Gravity 40-41
Group 1 98-99
Group 7 88-89
Growth 12-13

H
Halogens 88-89
Hazard symbols 102
Heart 25
Insulators 46-47, 51
Ionic bonding 81, 84-87

I
Ionic compounds 84-86, 91, 103
Ionic structures 86-87
Ions 54-55
Isotopes 21, 60-61, 66-67, 75

K
Kinetic energy 42-43

L
Mains electricity 58-59
Mass 13, 36-37, 114-115
Mass spectrometry 95
Meiosis 10-11
Metal Carbonates 108-109
Metal oxides 106-107
Metallic structure 96
Metals 96-96, 100-101
Minerals 22
Mitosis 10-11
Moles 112-113
Momentum 38-39, 41
Monomers 92-93
Mutations 11

N
Nanoparticles 91
Neutralisation 106-107
Noble gases 78
Nuclear decay 61
Nuclear equations 61
Nuclear power 68-69
Nuclei 4, 60

O
Organisation 4
Organs 4
Osmosis 14-15, 26
Oxidation 100, 111
Oxygen debt 17

P
Parallel circuits 50-55
Percentage composition 113
Percentage yield 115
Perfumes 93
Periodic table 75-79
pH scale 105
Phloem 26
Photosynthesis 20-21
Plastics 92-93
Polymerisation 92-93
Polymers 92-93
Potential difference 52-53
Power 44-45
Precipitation reactions 83, 107
Production processes 90
Proteins 7, 8-9

R
Radioactivity 62-67, 69
Radon 64-65
Rates of reaction 116-117
Reaction 40-41, 80-81, 89, 99, 101, 106-109, 116-121
Recoil 41
Reduction 100,111
Relative atomic mass 112
Relative formula mass 112-113
Resistance 37, 54-55, 59, 97
Resistors 54-57
Respiration 16-17
Resultant forces 36-37
Reversible reactions 118-119
Rf value 94
RNA 7
Rollercoasters 43

S
Salts 106-108
Sampling 18-19
Series circuits 50-55
Smart alloys 96-97
Sodium chloride 110-11
Specific heat capacity 121
Speed 32-35
Speed-time graphs 34-35
Stars 70-71
State symbols 83
Stem cells 12
Stopping distances 39, 44-45
Superconductors 97
Symbols 80
Synthesis 90

T
Thermoplastics 93
Thermosetting plastics 93
Tissues 5
Transition metals 79, 100-101
Transpiration 26-27
Transport 14-15, 24-27

V
Velocity 32-35
Voltage 52-53

W
Water 98
Water movement 26
Weight 36-37
Work 42-43

X
Xylem 26

Z
Zonation 19